Advice to Rocket Scientists
A Career Survival Guide for
Scientists and Engineers

robotics.
— space

Pooya's   1. sys-id
Prop.
STTR   2. RTOC on-chip
3. PS LTIE?
4. AutoQFT

redsing?
QUAV
4. Research for NL sys.

lin bnds   Topics define
to PhD?
everything.

Advice to Rocket Scientists

A Career Survival Guide
for Scientists and Engineers

by Jim Longuski, Ph.D.
Purdue University
West Lafayette, IN

American Institute of Aeronautics and Astronautics, Inc.
1801 Alexander Bell Drive
Reston, Virginia 20191

American Institute of Aeronautics and Astronautics, Inc., Reston, Virginia

4 5

Library of Congress Cataloging-in-Publication Data

Longuski, Jim.
  Advice to rocket scientists : a career survival guide for scientists and engineers / by Jim Longuski.
    p. cm.
Includes bibliographical references.
  ISBN 1-56347-655-X (softcover : alk. paper)
  1. Aerospace engineering--Vocational guidance. I. Title.

TL850.L66 2003
629.1'023--dc22

2003022104

Book design by Chris McKenzie

Cover design by Aaron Bertoglio

MATLAB® is a registered trademark of The MathWorks, Inc.

Table of contents

Preface

If you are a rocket scientist, want to become a rocket scientist, or know and care about a rocket scientist—this book is for you.

While rocket scientists are generally regarded as brilliant when dealing with space travel, physics, mathematics, and technology (that is, things), they are often notoriously inept when dealing with office politics, personality conflicts, and power struggles (that is, people). They are sometimes innocently ignorant of such things. Because of this blindness to the darker side of human beings, the rocket scientist can be hurt by the political struggles that go on in the aerospace industry.

This little book is concerned with the rocket scientist's happiness. It is a survival guide for rocket scientists. It tells you why the workplace is different from school; how to seek out enlightened managers; how to negotiate your first job offer; how to tell your boss, "we've got a problem"; how to give a presentation to rocket scientists; how to write a technical report; how to achieve visibility and why it is important; and how to become a professor of rocket science. In short, this book tells the rocket scientist not only how to survive, but how to be happy and how to flourish in the complex world of the aerospace industry where science and politics often clash.

There is no other book like it.

I wrote this book as an outgrowth of my lecture "What Your Professor Never Taught You." I have been advising colleagues and students on how to survive in the workplace of aerospace engineering and academia for nearly two decades. The stories I present here are real-life ones. The conclusions and observations, however, are my personal opinions and are necessarily subjective. While my way is but one interpretation, I can say with confidence that many of my former students are successful, happy aerospace engineers and professors, in part because of the advice I give here.

The principles in this book also apply to many fields and industries in which there are managers and highly trained analysts. This advice is sorely needed wherever there is a potential conflict between "political" reality and physical reality.

I dedicate this book to all such analysts who find themselves at the mercy of office, and even national, politics.

By the way, I'd like to explain how I chose to deal with the he/she pronoun dilemma. Some authors have found that simply alternating she and he neatly solves the problem. But the aerospace industry is dominated by males—and in this book I relate real-life stories that often involve dumb mistakes made by (guess who?) men. It would not be fair to attribute half the errors made in the field to women (who are the minority)! So in the interest of fairness and accuracy I have kept the blame on the male pronoun, whenever it applies. Throughout the book I address the reader as "you"—a pronoun which, fortunately, does not discriminate between the sexes. It is my sincere hope that "you" includes a great many female readers who aspire to be—or already are—rocket scientists.

Jim Longuski
October 2003

Acknowledgments

I'd like to thank all those who contributed to this book by their positive support, their helpful suggestions, and their sharp eyes for typos. Among these are my friend and colleague, Professor Tasos Lyrintzis; my graduate students, K. Joseph Chen, Theresa J. Debban, Kristin Gates, Wyatt R. Johnson, Damon F. Landau, T. Troy McConaghy, Masataka Okutsu, Steven G. Tragesser, and Chit Hong "Hippo" Yam; and my undergraduate students, Amanda Allen, Luca F. Bertuccelli, Eric Briggs, Chris Burnside, D.C. "Craig" Hamilton, and Nicholas Saadah.

I thank my mother, Jeanette T. Longuski; my brother, Joseph A. Longuski; and also thank my secretary, Sharon Wise, for typing the corrections.

Above all, I thank my wife and best friend, Holly C. Longuski, who has given me the greatest encouragement and support.

$$\vec{F} = m\,\overset{e^{\cdots}oc}{\vec{r}}$$

Part 1

Who should Read This Book

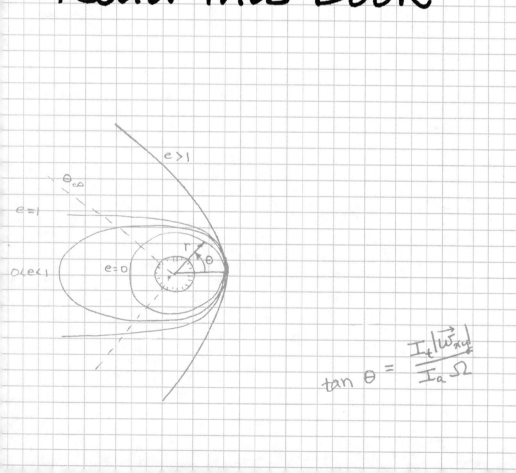

Chapter 1:
Who is a Rocket Scientist?

If you have a degree in aerospace engineering or in astronautics, you are a rocket scientist. If you work at NASA, JPL, APL, the Aerospace Corporation, or Draper Lab, you are a rocket scientist. If you work at Lockheed Martin, Boeing, TRW, or any other aerospace engineering company, you are a rocket scientist.

But are you the best rocket scientist that you can be?

Probably not.

This book will show you how you can be the best.

Chapter 2:
It Doesn't Take a Rocket Scientist to Be a Rocket Scientist

After yet another NASA failure, a late-night television comedian quipped that "apparently at NASA it doesn't take a rocket scientist to be a rocket scientist."

The term "rocket scientist" has become a part of the American language, on par with the terms "brain surgeon" or "genius."

But rocket scientists are not always measuring up to the country's expectations. Clearly it's time for rockets scientists to start living up to their name.

$$\vec{q_f} = \int_{t_1}^{t_2} \vec{F}\, dt = m\vec{V}^{\,e \to oc}\Big|_{t_1}^{t_2} = \sum_{i=1}^{z} m_i \, \vec{v}^{\,e \to OP_i}\Big|_{t_1}^{t_2}$$

Part II

Why the Boss is Key

$M_x = I_x \dot{\omega}_x + (I_z - I_y)\omega_y \omega_z$

$M_Y = I_Y \dot{\omega}_y + (I_x - I_z)\omega_z \omega_x$

$M_z = I_z \dot{\omega}_z + (I_y - I_x)\omega_x \omega_y$

Chapter 3:
It's Not About Grades

You've got your degree or degrees. You may have a Bachelors or Masters, even a Ph.D. You may have a 4.0 GPA, 1600 SATs, and 2400 GREs. You may have graduated first in your class at MIT, Caltech, Stanford, or one of the Ivy League schools.

So you should be a smashing success in the workplace, right?

Not necessarily. But you'll certainly be offered a job.

On the other hand, you may have less-than-perfect grades, SATs, or GREs and have gotten your single degree at one of the other schools. Should you chalk it up as a failed career?

No way! You will have more difficulty getting job offers, but after you start work you're on a whole new playing field.

Chapter 4:
Why the Workplace is
Different from School

I t comes back to the Peter Principle: those skills that promoted you from the last station finally become irrelevant in the next, and you've reached your level of incompetence. You have reached your final level and are no longer promotable. (See the original book, *The Peter Principle*, by Laurence J. Peter and Raymond Hull, for a profound and enlightening discussion of how hierarchies operate.)

Let's consider an example.

Back in the classroom, the professor asks a question: "What is the principle of rocket propulsion?"

A student raises his hand and says, "The exhaust gas pushes against the air to accelerate the rocket into space." The professor looks around for any other answers. The golden boy with the 4.0 casually raises his hand.

"Yes?" asks the professor.

"The answer just given is the popular misconception. If it were true, no rocket would work in space where there is no air to push against," says Golden Boy. "The correct answer is based on the linear momentum exchange. A particle of propellant is ejected as exhaust carrying a certain momentum. By the principle of action and reaction, the rocket gains the same momentum. As long as there is propellant to be ejected, the rocket will continue to accelerate, even in space. In fact, it is more efficient in space where there is no atmosphere to slow down the flow of exhaust gas."

The professor beams a smile of approval upon Golden Boy. Not only has he given the correct answer, but he has devastated the previous answer (and the student who gave it). Golden Boy has proven his superiority, and in the competitive environment of the classroom he has demonstrated that he is better than all the other students. He will, of course, receive an A.

Now let's consider what happens when Golden Boy gets into the workplace. He has only been at the prestigious laboratory for six months, but he has already assessed that he knows more than his immediate boss about spacecraft maneuver analysis. The boss of his boss of his boss has called an all-hands meeting for a critical review of mission plans, including spacecraft maneuvers. Golden Boy's boss will present an evaluation of the maneuver analysis, work mostly done by Golden Boy himself.

During his boss's presentation, Golden Boy's hand casually raises. "Yes?" asks his boss, nervously, "Do you have a question?"

"More of a comment rather than a question" says Golden Boy. "Your spacecraft maneuver analysis is incorrect because you assumed you'll have gyro readings available

---

to calibrate the accelerometers, but you forgot that in all-spin mode the gyros will be saturated, so your assumption is incorrect and your analysis is therefore invalid."

At this point, Golden Boy looks to his boss's bosses for a gleam of approval. He sees a wry grin on the top boss's face, what might be interpreted as anger on another's and in his own boss's face he sees confusion and embarrassment.

Later that year, Golden Boy gets a below-average raise.

What has he done wrong?

Golden Boy has forgotten about the Golden Rule or, more likely, has never heard of it.

Chapter 5:
The Golden Rule: Make Your Boss Look Good

The rules have changed and Golden Boy doesn't realize it. He thinks he's in competition with his boss, but in fact he's made a blunder of the first magnitude, because the Golden Rule is to make your boss look good.

Why must you make your boss look good? Because you are now in a cooperative venture with your boss, and with all the team members of your company.

It is your company or laboratory that is in competition with others. Competing with your boss is detrimental to your organization. Keep in mind that you are working in a hierarchy and that one of the rules of such hierarchies is that you will never be promoted above your boss. (The exception is to make a lateral move to another section or company as discussed in Chapter 26.)

For you to get promoted, your boss has to be promoted first. Then, if you are his right-hand man, it will be necessary for him to promote you so that he can continue to benefit from your talent, energy, and loyalty.

Chapter 6:
Does This Mean You Have to Kiss Butt?

Of course not! Only incompetent bosses like their butts kissed. And you don't want to work for them anyway (more on this later).

Does it mean you have to agree with your boss, even when he's wrong? Most emphatically, no! The competent boss doesn't want a "yes" man. He wants a competent engineer who will give him accurate analysis in a timely manner. He wants to be fully apprised of your analysis at all times, especially when it's in disagreement with his ideas. Above all, he wants to hear bad news early, so he can solve the problem before it's too late.

Your boss has hired you to solve problems. He has entrusted you with important work and wants clear, accurate answers from you. After a while you will know more about spacecraft maneuver analysis than he does. You know that and he knows that.

But you are only working on spacecraft maneuvers, and your boss has responsibilities that include a whole raft of other problems (such as scan platform pointing, attitude determination, gyro and accelerometer calibration, star identification, attitude control and articulation, and so on.) Your boss has Ph.D.s working on these problems, and he has to stay on top of them. He needs to maintain the big picture. He also has to maintain a budget and often has to fight with higher-level management to keep his employees (including you) on the payroll.

He has a lot of responsibilities and doesn't need his latest hire upstaging him during the critical design review.

So, what if your boss is wrong in his understanding of your work on maneuver analysis? Then you tell him. But not at the critical design review! You tell him behind closed doors.

Your boss will inform you ahead of time that he has to make a presentation to his bosses and he needs your input and maybe some slides with nice pictures of the spacecraft and a few numbers indicating the performance capabilities you have calculated.

You make the best slides you can—models of clarity and accuracy (more on this later). You make an appointment with your boss to go over the slides. You correct any misunderstandings he has in the privacy of his office.

You explain everything clearly and accurately. You do not allow your boss to make a mistake in public.

His mistake is your mistake. If your boss doesn't move up in responsibility, then neither will you. You're a team.

Chapter 7:
What if My Boss
is Incompetent?

He isn't because you would never work for an incompetent boss. This is because you checked him out before you accepted his job offer. (How you did this will be explained later.)

Which brings us to this pop quiz: which is the most important factor in your job?

A. the salary
B. the work
C. the boss

The correct answer is C, the boss. Why? Because if the boss is incompetent (that is, lacking vision, drive, and/or integrity) then it won't matter how interesting the work is or how much you're getting paid.

A fundamental rule of how hierarchies work is that high-level management always takes the word of lower-level managers over the word of employees.

If your boss is incompetent and he justifies your low salary raise by telling upper management that your work is somehow lacking, then that assessment is taken as gospel—your work is lacking. Your boss feels sorry for you, so he keeps you on the job in spite of your lackluster performance. Your objections to his boss will be taken with a grain of salt: the protestations of a disgruntled (lackluster) employee.

Let's consider an example.

A young man with excellent grades from a top-flight aerospace engineering school joined the navigation section of an important government laboratory. He was excited about the work and very happy with his starting salary. His first boss, a stocky, handsome man—who looked a lot like Orson Wells in his latter years—seemed OK.

The young man assumed his boss was like one of his professors. The boss had good credentials and his publications had been referenced in important books on trajectory optimization. The young man looked up to his boss for guidance in his career at the laboratory. He assumed his boss cared about his career and advancement.

The young man wanted to learn everything he could about how to navigate spacecraft among the planets. "I want to study the problem, derive the equations, and check the lab's simulator for accuracy," he told his boss.

The boss shook his head. "I don't want you to reinvent the wheel here, boy! Look, I'll tell you what to do. You put these numbers into the NAV simulator and bring the output back to me and I'll tell you what it all means."

Disappointed, but still hopeful, the young man did as he was told. "Maybe later," he told himself. But after a year he found that the boss would never allow him to investigate or study the problem of space navigation in the way he had learned at school.

The boss never let him spend time on understanding the underlying issues.

After the first year the young man should have left that job, but he didn't. His first mistake was to accept the job at all, but he went on to make a second, even more serious, mistake: he stayed with that boss for another six years. The young man was afraid to change, afraid to appear disloyal even after his boss demonstrated that he didn't deserve loyalty.

Every year the young man got lower and lower salary raises.

He finally tried to find another job. Luckily, he had a very good interview at another top government laboratory. The immediate bosses liked him and passed his resume all the way up to the vice president, who approved all new hires.

Then the young man's dreams were dashed because the vice president did not give approval. Why not? The young man's salary was too low. "It's a sure sign of a problem when a candidate from a top lab has a low salary like this," the vice president said.

Chapter 8:
Check Out Your Boss
Before You Accept the Job

So now you can see why it is crucial to check out your boss before you agree to work for him.

But how do you do that? During your interview, do you ask him point blank: "Are you a good boss? Competent? Trustworthy?" Or, when you meet your boss's boss, do you ask him those questions of your prospective supervisor?

Clearly, you cannot.

However, there are two ways to investigate the boss and you should use both of them.

The first involves the interview with the boss himself. Most interviewees think that they are the subject and should be on their toes to give the very best impression. But they should also be interviewing the boss and he should come across showing his best behavior. Both individuals are (or should be) trying to positively impress each other.

What are you looking for in a boss? A person with vision, energy, intelligence, honesty, fairness, and value for human beings. Few bosses will have great strengths in all of these categories, but a severe weakness in any spells trouble.

You can observe many of these qualities during your interview. You need to be a keen observer and, most of all, you must trust your gut instinct.

If you thought the boss was highly intelligent and visionary, but you wondered about that remark he made about your hairstyle, your manner of dress, your speech, your ethnic or racial group, or your sex (or sexual orientation) then you need to think very carefully about that remark. Make no excuses for the boss. Remember, he's on his best behavior and if he has let an off-color remark slip, how is he going to treat you when you're vulnerable to his control and assessment? Multiply his remark by a factor of ten and that's what you'll likely get in his way of treating you. (In some cases, you should probably consult a lawyer, but such problems are beyond the scope of this book.)

Do not work for a boss who makes any such remark during an interview.

The second opportunity (to investigate your prospective boss) occurs when you are introduced to his immediate employees. You will, of course, be anxious to know as much about the technical details of the job as possible; now is your chance to ask those questions. Ask away. Don't be surprised, when the work requires a security clearance, if you do not get very specific answers about the nature of the tasks.

But the boss's character is no secret and is of prime concern. Bring up your questions about the boss during your one-on-one interviews with the employees. Do it in private. The employees will be your colleagues and you should consider them your equals. Close the door (if there is a door). Do not talk to a group of employees about the boss. No one will give you the information you need in front of others. It must be private. It must be one on one.

"So what do you think of the boss?" you might ask. If the employee is hesitant he or she might say, "What do you mean?" Throw it back to him: "Do you like working for him? Are you happy with your raises? Is he a good leader?"

These are fair questions. You may be surprised how quickly the answers spill out, especially if they are bad news. Bad news about a boss travels fast, and is easy to obtain. Of course you have to weigh the possibility that you might be talking to one of those lackluster employees. So be sure to get several opinions.

A consensus will quickly build. If the boss is OK you will be told "he's OK." If he's a rising star and great motivator, employees may sing his praises, but not necessarily. Rocket scientists are reluctant to praise anyone too highly, so don't expect nominations for the Nobel Prize or for canonization to sainthood.

But if the boss is bad, you will hear about it! "The only calculator he knows how to use is a four banger." (Add, subtract, divide, and multiply.) "He's been a supervisor for 25 years and won't be moving up in the next 25." "He's a salesman, but no analyst, the upper management likes him because he brings in government funding, but he's awful to work for. A little tin-pot dictator with Napoleonic delusions." "He tried to publish my work in a journal when I was on vacation. He gave himself top billing and screwed up all of my equations." "He doesn't know his periapsis from his semi-latus rectum." "He's an egocentric anomaly."

Yes, bad news travels fast indeed, and you don't need to dig deep to hear the juicy stories.

It will be disheartening to hear them, after the exciting plant trip, the challenging, interesting work, the anticipation of an excellent salary at a prestigious laboratory. But you must hear them and make your decision accordingly. You must protect your career from the disaster that will surely follow if you accept a job with an incompetent boss.

$$\tan \gamma = \frac{|\vec{\omega}_{xy}|}{\Omega}$$

Part III

How to Get the Right Job

$$-\frac{\mu m}{r^2}\hat{e}_r = -\frac{\mu m}{r^2}\cos\gamma\,\hat{u} - \frac{\mu m}{r^2}\sin\gamma\,\hat{v}$$

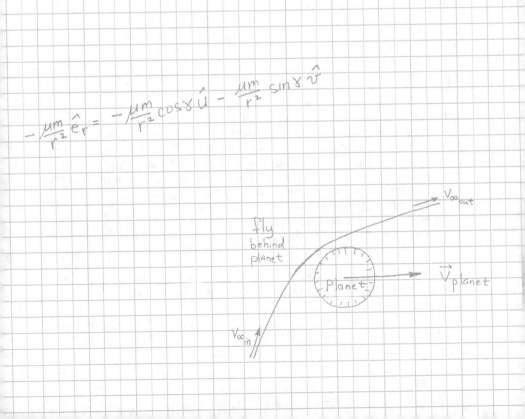

Chapter 9:
Why You Need Two Resumes

The placement office at your school or university and the human resources (HR) departments in the aerospace industry like to cut down on their paperwork. They want you to fit your resume on a single page. They tell you that two pages is not acceptable if you want your resume maintained in their files. This is understandable: they have thousands of resumes to deal with.

For some reason students have gotten the idea that the one-page resume is the ideal resume, ordained by a higher authority, but it is merely a paper-saving policy of the personnel services. So the student will cram all his credentials into one page and ship out his resume. Usually he doesn't think he has much more to say anyway: "I have a Bachelor's in aerospace engineering and a 3.0 GPA—what's the big deal? What else can I say?"

You can say plenty and take plenty of pages to say it — that is, if you want your dream job and you want to catch the attention of the supervisor who has it on tap. "Wait a minute, I should send my resume to a supervisor? Won't he see it on file in the HR department?"

Yes, you want to send it directly to the supervisor and no, he won't find it at HR. The reason he won't see it at HR is that he's a very busy person. He's short-handed right now and so he's working overtime to make up for the shortage. He desperately needs a launch vehicle analyst and he doesn't have time to go to HR. "Besides, have you seen the files at HR? There are thousands of resumes down there and they're not organized according to job description, they're organized by degree. It's like looking for a needle in a haystack." The supervisor hopes his phone will ring and you will be telling him how much you want to work on launch vehicles.

The supervisor told his employees to keep an eye out for a new hire. They don't need to be reminded. They're really pressed for time. The preliminary design review is right around the corner and they need to make a decision about which launch system to propose. The supervisor also informed his bosses of his need to fill the position. Everyone's on the lookout for a launch vehicle analyst.

Now, you've spent the last two years polishing up your Master's thesis on design of gravity-assist trajectories for exploration of the solar system. It's not about launch vehicles.

On your one-page resume you also show a Bachelor's degree in astronomy and a four-year stint at Rockwell. You didn't bother to mention in your one page resume that you worked on assessments of future Shuttle-derived configurations (launch vehicle!). You did lots of things at Rockwell and there was not enough room to mention everything.

So, even if HR was on its toes (it is, in principle, possible) they wouldn't have pulled your file and sent it to the supervisor. There are hundreds of aerospace engineering Master's degrees on file and yours doesn't say "launch vehicle analysis" anywhere on it. You are never contacted.

Now let's consider an alternate future. (Keep in mind that there are no rigid rules for the long resume. The example that follows will give you some ideas you can use or alter to suit your needs.)

You have carefully put together a five-page resume that provides the details of your academic and work experience. On the first page you list your name and every conceivable way you can be contacted—school address, local address, permanent address, voice and fax numbers at each address, e-mail. You may wish to state a goal, but it is not necessary. If you want to give an objective, make it as general as possible or make a list of possible career directions. Don't worry about consistency in your goals list, but do keep consistent with your passions and interests and be yourself. Rocket scientists are often too analytical and hence abhor logical inconsistency. Put those logical objections aside and write:

"I'm looking for a highly technical position to capitalize on my specialization in gravity-assist trajectory design, but I also have an interest in reentry, control, attitude dynamics, optimization in aerospace engineering, propulsion, and pursuing a management position."

Even if you're a Vulcan you should appreciate that the supervisor who examines your resume will be looking for some specific characteristics. If you list only one specialty, then your probability of landing a job will be small. On the other hand, if you use the shotgun approach of making a long list, you increase your odds considerably. Be flexible. There are a lot of interesting jobs out there that you are qualified to do but know nothing about. Trajectory analyst? Yes. But management track? Why not? Management could use a few more Mr. Spocks (or even one Mr. Spock). Chess is a management game and the more logical manager wins (with exactly the same personnel)!

On the first page also list your degrees and year awarded (starting from most recent: e.g. Ph.D. 2000, M.S. 1994, B.S. 1992). If you graduated with honors, include them (B.S., cum laude, 1992). Give the university and department.

Do not give your GPA unless it is exceedingly high (3.99/4.00). Listing a 2.87/4.00 GPA hurts you on the resume. (Of course, this information is available on your transcript, which every competent supervisor will request. Some companies require that the GPA be listed in the resume, in which case you must comply or risk having your resume rejected.) You want to sell the company on your knowledge and experience. If you are a good match to the company's needs, then your less-than-perfect GPA will be ignored when the supervisor finally looks at the transcript.

List any employment experience that may be relevant. If as a student you only had a job working as assistant manager at McDonald's, you should list it because it demonstrates that you were capable of holding a job and even took a responsible position. (You showed up for work every day!) If you have a higher degree (Ph.D. or Master's) then don't list the McDonald's job. You will have had other experiences that are more important such as research assistant, teaching assistant, laboratory assistant, or a fellowship.

On your first page, give a brief description of your academic background. With a Bachelor's degree you might say that you have a Bachelor's in astronautical engineering with a major in astrodynamics and attitude control and a minor in rocket propulsion. Courses included orbit mechanics, linear and nonlinear control, Lagrangian dynamics, attitude estimation, space systems design, thermodynamics, fluid dynamics, trajectory optimization, solid and liquid chemical rocket propulsion systems, and internal ballistics. Mathematics courses included differential equations, linear algebra, complex variables and numerical methods. You have a working knowledge of MATLAB®, Fortran, Unix, C, and C++.

You see? You have a lot to say about yourself, even with just a Bachelor's degree.

If you have written a Master's thesis, you should write a paragraph describing the work. The abstract of your thesis will do nicely. If you wrote a Ph.D. dissertation, then write a paragraph on it as well (put this paragraph above that of the Master's).

Starting with the second page, you can provide details on your work experience (if any). "Managed five people at McDonald's. Made work schedules, trained employees, operated cash register." If you worked as a teaching assistant, describe all your duties: "Graded homework and exams, tabulated grades, held recitation class, occasionally lectured in professor's absence, gave evening help sessions." If you obtained a good evaluation from the class, indicate it: "Received a 3.0/4.0 score on class evaluation question: this is among the best teaching assistants I have ever had."

Of course you should mention all honors and scholarships you have received. If you won the best teaching assistant award in the astronautics department or the Shuttle memorial fellowship you should definitely list them in your honors and awards section.

But what if your have no honors or awards?

Go on to list the titles of the major courses of your program. Remember on page one where you mentioned that course on trajectory optimization? Now you list that title and go to the course catalog and type the description (updating according to what you were actually taught. Most course descriptions are sadly out of date and you want to make sure you've included the exciting new stuff the professor lectured on.)

Do this with all the major, minor, math, and computer courses. You will end up with a few pages of course descriptions. It's a lot of work.

So why should you do it?

Because you want the best, most rewarding career opportunities that a person with your education deserves. You should not assume that all aerospace engineering academic programs are the same. In fact, even your colleagues from the same program will have a different list of courses on their resumes. And if they didn't read this book they won't even think of making such a detailed list.

The detailed curriculum list provides a sort of Rorschach test for the supervisor who scans it. The Rorschach test consists of meaningless inkblots that psychologists show to patients. The psychologist would ask patients what they saw in the pictures.

Q: "What do you see in this picture?"
A: "A man and a woman facing each other."
Q: "How about this one?"
A: "A man and a woman kissing."
Q: "And this?"
A: "A man and a woman embracing."
Q: "And how about this one?"
A: "A man and a woman lying down together."
Q: "You sure seem to be obsessed with sex."
A: "Hey, you're the one with the dirty pictures."

Let's say your prospective supervisor is obsessed with finding a launch vehicle specialist. When he sees that you've taken courses in orbit mechanics, trajectory optimization, and differential equations, not to mention the work you did on Shuttle-derived launch vehicles for Rockwell, he will see a launch vehicle analyst in you. And you never knew you were a launch vehicle analyst.

On the other hand, another supervisor will see an entirely different you in your resume. That's because you've taken a fairly general program in aerospace engineering. But in addition, from your detailed course descriptions, the supervisor sees specific details that he puts together in a package that fits his particular needs. Bingo! You're the one.

Does he care about all the other courses you've taken? No! He doesn't even notice them.

This multiple-page resume packs a wallop. It is a powerful technique for getting the right kind of job offers—but nobody knows about it. Except you—you've just read it.

The next question is, how do you get your resume to the supervisor in the first place?

Chapter 10:
Getting Your Resume
to the Right Person

The boss is looking for someone like you. You've sent your resume to HR and given it to the placement service at school. So you've done all you can, right?

Wrong. The boss will never see your resume at HR.

You've done a lot of work putting together your long resume. Now you've got to campaign for yourself. You've got to use the telephone. Call some companies. Get referrals to the engineers who do the real work. Forget about HR. Forget about the placement office. (File with them and forget them.) Call some people!

Aerospace engineers are great people to talk with. After you make contact with one engineer in the company, you can readily connect with others. It's easy. Find a phone number, any number. Call.

"Hello, this is Joe Student. I will graduate with my B.S. in A.E. next month and I'm looking for employment at TRW."

"What kind of work?" the voice at TRW asks.

"Astrodynamics, spacecraft control, or propulsion."

"Well this is the structures department. That work is done in the guidance and control section."

Don't worry, ask for a referral.

"Could you please give me the phone number of someone in the guidance and control section?"

With 99% probability you will get a reply something like

"Hold on, Joe, let me look it up for you."

Thank the person and make your next call. For some reason, many students are afraid to make phone calls. What do you expect to hear?

"Who the hell do you think you are to call here?"

It won't happen. Rocket scientists are some of the finest people in the world and they want more good people to join their ranks. Call the guidance and control section.

"May I speak to the manager?"

Be polite. The manager will give you the straight story. If he has the slightest opportunity to hire you he will ask for your resume. Even if he doesn't have an opening he will ask for your resume. If you are perfect match to his needs (or have a stellar resume), he will create a position just to hire you. Positions are made for top candidates.

Write down the manager's address. Make sure you get his title right. If he has a Ph.D., you should address the letter accordingly.

Dear Dr. Control,

After discussing employment opportunities at TRW with you today (give date), you encouraged me to send my resume to you. (Please see enclosed.) I am looking forward to visiting you for further discussion about a career in your guidance and control section at TRW.

Sincerely,
Joe Student

And while we're on it, let's talk about letter writing. Whatever happened to "dear" and "sincerely?" These are the standard opening and closing statements of any formal letter. They show good manners and respect. Forget about using "To: Dr. Control;" "Hello, Dr. Control!;" "Hey, Dr. Control!" and any other replacements for "dear." It's common courtesy. It doesn't mean you're writing a love letter or marriage proposal (use "my dearest" for that!).

Once you've established contact with a potential employer, you can take the opportunity to restructure your long resume to emphasize your strengths that are of special interest to the company. Before sending out your cover letter and resume, make sure they have no typing or grammatical errors. Have a friend or advisor proofread both documents. A great way to proofread for yourself is to read everything out loud. You'll catch poor sentence structure and typos with astounding accuracy this way.

Make your letter and resume perfect, then send it out within a day of your telephone conversation. After five days call the manager again to make sure he received it. If he hasn't, send another immediately. If he has received it, you will have an opportunity to get his reactions to your resume. Do not talk for long. The boss is busy. Take notes of your conversation.

"Hello, Dr. Control? This is Joe Student, have you received my letter of January 1?"

If the answer is yes, you can take one shot at probing:

"Based on what you see in my resume, do you think there is an opportunity for me at TRW?"

Whatever he says, be polite. Follow up on his instructions, which you have written down. Maintain contact. Leave nothing to chance.

Your letter doesn't get there. You call and find out. You send a second one. The secretary says the boss will call. Call the boss back if you don't hear from him within two days.

Maybe the secretary forgot. Maybe the voicemail system broke down.

Make the call.

"Won't they think I'm a pest?"

No. They will see your persistence as a virtue. As desire, commitment. As someone who knows what he wants.

Rocket scientists like that in their candidate employees. Don't forget that rocket scientists constantly deal with triply and quadruply redundant aerospace systems. "If something can go wrong, it will!"

When you persist, when you don't leave anything to chance, when you don't allow anything to go wrong, they see you as one of them. You have distinguished yourself among all other job applicants. They will be favorably impressed.

Chapter 11:
What About References?

Your references should be on the last page of your resume. You should list about four people who can comment about your suitability for employment.

You should start thinking about who to list when you begin college, not when you are about to graduate. Cultivate relationships with professors and employers who have been impressed by you. Be very selective about your references.

Should you use the professor with the most distinguished credentials? What about that Nobel laureate in that physics class you had (along with 300 other students)? And what about Auntie Christina who thinks you are just wonderful in every way?

These are examples of extremes you want to avoid.

It's no use to go for a big name if that person doesn't recognize your name when a prospective boss calls. So unless the Nobel-prize-winning physicist was so impressed with you that he called you by name and was enthusiastic when you asked him to be your reference, don't list him.

Having a reference say "Who? No, I don't know him. Sorry," is devastating to any further consideration of your application.

Auntie Christina's glowing remarks about what a bright boy or girl you were in the first grade science fair won't cut it either.

Strive for a balance. You want someone with strong credentials who will give you a strong recommendation.

A good way to ask (and you must always ask for permission to use someone as a reference) is to say "Professor Somewhat Well-known, do you feel you know me well enough to give me a strong recommendation for a career in aerospace engineering?"

If Professor Somewhat is doubtful, you've given him a way to save face. If he suggests that one of the professors you worked for as a teaching assistant would have more to say, then bail out of his office as soon as you can. (Be sure to thank him.)

Pick someone who is expressive and optimistic. You want a "bubbly recommendation." Someone who will say, "He's top notch," not "he's OK."

Of course if you goofed off all four years in college and insulted every professor and boss you ever knew, you are in serious trouble. The aerospace business is a small world: treat everyone well—it's a good philosophy and good common sense. Word gets around and a bad reputation is nearly impossible to correct.

Make sure your references have a copy of your resume so they can refer to it when they get the call. You might even ask for feedback on your resume from your references. You'll probably get positive comments on your long resume and maybe a few good tips or corrections to typos.

List every conceivable way that your references may be contacted (ask for permission to list home phone numbers.) You want the boss to be able to contact your references ASAP (as soon as possible). He's in a hurry.

Don't get cute with a line like, "References provided upon request." Your prospective boss doesn't have time for that. Make it as easy as possible for him to hire you.

By the way, one more advantage of your long resume is that it provides your references with many extra details that they can mention when they are asked to write a letter of recommendation or comment on you in a telephone interview. Your outstanding, unique background sets you apart from the crowd. You've made it easy for your reference to point this out.

Chapter 12: What to Bring to the Interview

Be dressed and well groomed.

Wear a smile.

Present an enthusiastic, inquiring attitude.

Speak clearly.

Listen well.

Have impeccable manners.

Be yourself.

These things are obvious, but what else can you do? What should you bring to the interview besides a winning personality and good personal hygiene?

Documents.

You should bring at least three documents. They are

1. your long resume,
2. your transcripts, and
3. your senior design project, and any other important reports or papers you have written.

These include projects you have done for your engineering courses, conference papers you may have coauthored with your advisor, and any journal publications.

If you have a Bachelor's degree in aerospace engineering you will have three documents. If you have advanced degrees, you will have more.

Bring 15 stapled copies of your resume. Bring an original loose copy of your resume. Bring 15 copies of your transcripts plus the original. Bring the bound senior design report plus a loose original copy.

When you find out who the head secretary (or administrative assistant) is, present that person with your original documents. The head secretary will then be able to provide copies to anyone in the company who wants them. The head secretary is a well-organized person who will not lose your documents.

When you interview various bosses and engineers, you should ask them if they have a copy of your resume. In theory they should have a copy because you sent them a copy through the mail. But here's what invariably happens:

"Did your get a copy of my resume?"

The interviewer looks around at the piles of papers on his desk.

"I think I saw it ... but I seem to have misplaced it ... "

Before he starts an archeological dig, you say

"I brought an extra copy for you. Here it is."

The interviewer will thank you. More importantly, he will have a written document in front of him, reminding him of your wonderful credentials and upon which he can take notes. When you leave, you will have left more than a lasting impression of your visit: you will have left documentation of it.

At an appropriate time during the interview, show the interviewer your senior design report. Mention to him that if he'd like a copy, you've left an original with the head secretary.

If your grades are stellar, you should offer a copy of your transcripts. If not, then wait to see if they are requested. Remember that your long resume gives a better picture of your talents than a less-than-perfect transcript. If you are asked for your transcripts, then hand them over. Don't make any excuses. Don't say, "Well, I didn't get the best of grades." Just say, "Oh yes, I've brought a copy for you." Hand them over without reservations.

When you go to the next interview, do the same.

You will most likely interview with half a dozen to a dozen people. They will all need a copy of your resume. That's why you brought 15. If you need more, the head secretary will make them for you.

Chapter 13:
Seek Out Enlightened
Managers

Enlightened managers make sure they hire the right person for the right job. Enlightened managers are matchmakers. They match not only your skills, but your desires, hopes, and dreams with the position you will find most fulfilling. Enlightened managers know that if the match is not good, they will be either forced to hire a new person in the next six months or will have to contend with an unhappy, unenthusiastic, noncreative employee for (perhaps) years to come. Enlightened managers know that rocket scientists have dreams—often inspired by science fiction—to travel the stars. Enlightened managers want to harness those dreams for the good of the company, the employee, and themselves. Enlightened managers recognize themselves in their potential employees.

Enlightened managers know more about you and your ability to do the work than you do. They've been in the business of hiring successful employees. They know how to read the signs; they know not only where you belong but where you'll be happy. They accurately assess your potential and your desires the way a good real estate agent matches the needs and buying potential of the would-be homeowner.

The good real estate agent knows (better than you) what you can afford. The bank will not let you borrow more than you can pay. You can trust their assessment.

You can trust the assessment of an enlightened manager. Do not be afraid if the job described to you sounds formidable, if you don't know how to solve the problem, or even what the problem is.

The enlightened manager has carefully studied your resume, transcripts, and reports. He has contacted your references. He knows your school and your professors. He has a challenging, stimulating job for you in which you can flourish.

Do not think about taking a lesser job, an easier job, a job you know how to do before you start it. That job will become very boring to you in the first year and intolerable after that.

The enlightened manager knows that you will need an interesting, difficult task. Something you can grow into. Something the company needs solved. Be confident in your abilities. You have what it takes. You have your degree and a solid foundation in the field of rocket science. You have learned much.

What you must realize is that there is no end to learning. The company expects you to learn many new things. A rocket scientist by nature is a lifelong learner. Only by learning new things can you expect to accomplish something. Only by rising to the challenges can you feel pride in achievement.

The enlightened manager knows this. He is a mentor. You already know that learning is hard work; it is also the only road to success as a rocket scientist. When your enlightened manager offers you a difficult, challenging job, you should be grateful. Accept his offer and count your blessings.

Do not be afraid.

Chapter 14: How to Negotiate Your First Job Offer

You've had wonderful interviews at several companies and the offers are starting to come in. You feel exalted by the possibilities, by all the things you might do—great salary offers, challenging work with opportunities for promotion and even higher salary, and good people to work with. It finally dawns on you that you must choose one and give up on all the others. You must move from the potential futures of many worlds to the actual selection of a specific career path.

The decision makes your gut ache. You were on top of the world, but now you feel sad, perhaps a little depressed. Don't be discouraged—most students go through a similar phase.

So how do you make that big decision? First, you must know yourself. You must know what you want. You must know both intellectually and emotionally what is best for you. You must know what feels right. Trust your instincts. Ask yourself, "What is my dream job?" Then listen carefully to your own answer.

Meditate on it.

How did you feel about the management? Your immediate supervisor? You don't know who they are? Then you must go back to Chapter 7 of this book and review why your boss is the most important factor in the job.

If you've made such a glaring error as to ignore who your boss is, you may still be able to recover. Call the company and find out. Then call the boss and go over a few things with him. Keep yourself on "high receive" when you listen to your boss—you are deciding on whether you can entrust your career to him!

If your prospective boss is irritated by your questions, isn't straightforward with you, or is short with you, then you can make your decision not to work for him. In fact, if the boss is anything less than delighted to talk with you further about your responsibilities, your benefits, your salary, your office, or your computer, don't accept the job offer.

But let's say that all the bosses were great and forthcoming about everything. Then you can concentrate on the job itself, which is next on the priority list. Take the most challenging, interesting, stimulating job. Take the job that leads to career, not just to money. Do not be afraid of taking on a difficult task—you can do it, and it will be more fun than you can imagine. Take the job that gives you the opportunity to work with the best people, the people you felt best about when you met them at the plant trip. The people you most admire.

If you think about these things carefully, you will know which offer you should accept.

So what do you do if your dream job with the great boss offers you the lowest salary? The answer depends on how low the offer is. You can easily find out what you are worth. The Society of Professional Engineering publishes salaries based on degree, years of experience, and location in the country. Your other job offers give you an idea of what you're worth. Your school's placement office tracks salary offers the students have received and posts them.

There is a going rate for a fresh B.S. in aerospace engineering starting in Los Angeles, for example. You should know the going rate. If your offer is significantly below the going rate, call your prospective supervisor and inform him that it is much too low for you to accept. If the going rate is $50,000 per year and you are offered $35,000 per year to work in Los Angeles, then you should not accept the offer.

Ever.

Not even if the aerospace industry is down and you can't get another offer?

Not even then.

Not even if my professors tell me that it's a buyer's market and I have to take what I can get?

Do you think your professors would follow their own advice?

So what do you do if it's your only offer? Enroll for another semester of classes. Take graduate courses. Take a Master's degree, a Ph.D., a postdoc. If working as a rocket scientist is your dream, then prepare yourself for an even better career when the market comes back up. If the dream has turned into a nightmare and you want to bail out, then take some courses in another field such as computing, business, law, medicine, or brain surgery. There are plenty of other careers a rocket scientist can pursue and flourish in. After all, you are a rocket scientist!

What if the offer is just a little low? Then what? Then you call your supervisor and say: "Boss, I love the work your group is doing and I'd sure like to work for you. In fact you're my first choice, but for one thing ...." The boss will say something like, "And what is that?" You reply: "Well, I feel a little stupid accepting a great job that offers less pay than two other offers I got. Is there any way you can make me feel a little smarter?" Or you can say: " I was hoping you could sweeten your offer a bit. I received two offers that were a bit higher than yours. But I'd rather work for you."

Whichever you say, the enlightened boss will ask you how much more the offers were. Be absolutely honest. If you picked a smart boss, he may ask you to fax a copy of your offers. If he doesn't, then you should make the suggestion. "How about if I fax my other offers to you and see what you can do about it?"

Your enlightened would-be boss will be happy to receive those faxes. Remember, your boss does not have the final word about salary. Salary decisions are made at a higher level. Your boss will have to make a case to his boss to increase your offer. He will need all the ammunition he can get. You have impressed him with your long resume, an excellent interview, your persistent interest, good grades, and strong offers from other companies.

Your boss wants to hire you. He doesn't want to lose you. Don't expect him to give you an answer over the phone. You have succeeded if he says, "Yes, I'll look into it."

Above all, do not make any ultimatums. Do not take any position other than that of being flexible and open. The biggest mistake you can make in negotiations is to kill the conversation. Always leave room for your boss to say no to your suggestion, without saying no to you.

This is a delicate discussion, and one that you haven't trained for. Rocket scientists are not born negotiators or politicians. They often have difficulty debating emotional issues. So you may feel uncomfortable asking for a higher salary. Keep a friendly tone and have a sense of humor about it (if possible). Your boss will do the best he can for you. If he comes up empty handed, you're still going to take the job and you want to be able to save your face and your boss's face when you do.

Suppose the boss comes back with a salary that matches your best offer. Should you go to the other company and get them to top it?

Stop right there!

The worst thing you can do is to price yourself out of your dream. One salary increase is enough. If you're getting the going rate or better from your dream job, then take it and be happy. Why do you want to know that the other company was willing to go much higher? Why spoil the result you worked so hard for: the best boss, the best job, and the best salary?

There is one other point over which you might want to negotiate, and that is the specific task you'd be working on. While you were interviewing at the company, you learned that there were several tasks you might choose to work on. You investigated this carefully by asking the employees, "What is the hottest job you've got?" You found out about the exciting new space system they were starting to work on. And you want to work on it, too.

When you have your final negotiation with your boss, tell him, "Thanks for the salary increase, now I can say categorically that I will accept your offer provided that you put me on Space Systems X."

Your boss will put you on Space Systems X even if he has to move someone off the job to get it for you.

Do all these things and you will land your dream job. You will be on your way to a great career in aerospace engineering.

$$\Delta\phi_{NR} \equiv 2\sin^{-1}\left(\frac{1}{1+x}\right)$$

Part IV

How to Survive in Industry

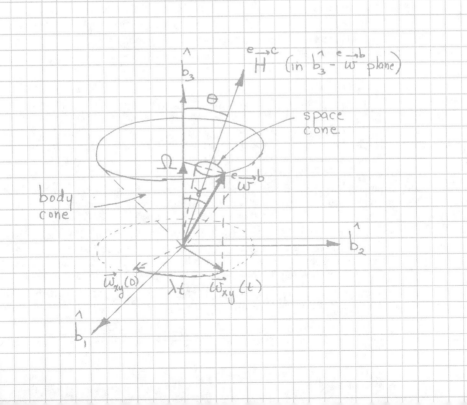

Chapter 15:
How to Survive Your
First Two Weeks on the Job

Your first two weeks on your new job will be the most difficult. You've moved far away from friends and family. You're in debt and your first paycheck won't come for two to four weeks.

On the other hand, your rate of pay is phenomenal compared to what you made as a teaching assistant or in work study at school. You're making five times as much money, you're working only half the hours you did at school, and you are contributing nothing, virtually nothing to the company.

You may start to feel a little guilty. Maybe you oversold yourself, misrepresented what you can do. All around you people are very busy accomplishing something. They are speaking a language you don't understand. There are a lot of acronyms being bandied about and you can't make any sense of it. "Why didn't I take that job at the IDTARS company?" you ask yourself. "I understood what they were doing."

You're scared and confused and feeling worthless.

Don't worry about it. Everybody goes through a stage like this.

Your boss comes into your office with a stack of papers 2 feet high. "A little reading material for you." he says. "These are the engineering memorandums, reports, and critical design presentations for Space Systems X. Read them ASAP."

"ASAP?" you ask.

"As soon as possible," he says.

Your boss leaves you with the stack. You pick the first report off the top and start reading. More acronyms. No definitions. Suddenly you feel very tired, but you keep slogging through the document. Occasionally you recognize a word or phrase, but for the most part it's a foreign tongue. You try to suppress a yawn, unsuccessfully. You are very drowsy, so you put your head on the desk for a moment. You rest your weary eyes.

You wake with a start. Did you hear your boss's footsteps? You look at your watch and 20 minutes have elapsed. You've been sleeping on the job for 20 minutes. You don't want to think about it but you just wasted one third of an hour, and you're making a good rate of pay.

You've read one page, which you didn't understand and you've taken a 20-minute nap. How's that for progress? You look at the stack of design specifications and a wave of depression moves through you. "I can't possibly read all this stuff," you say to yourself, and you feel like quitting. Twenty minutes on the job and you want to run away.

Don't be surprised if it happens to you. There will be plenty of documentation the company has accumulated over the last five years on Space Systems X, and it's complex material and no one can read it all.

It is impossible.

You are right about that.

So how do you learn about your job? How do you get started? How do you stay awake?

Reading company documentation on a rocket system is a lot like reading computer manuals to learn about how to run new software. Everyone knows how user unfriendly those user's guides are.

How do you learn to use the computer? You ask someone who knows. That's the fastest most efficient way to learn a new program.

So how do you learn to be a professional rocket scientist? Ask a rocket scientist.

Now go back to that stack of company documents. You thought you should read the contents. What you should be reading, instead, are the names of the authors. Look at the org chart (the organizational listing of all the employees in your section, organized by areas such as attitude dynamics and control, orbital mechanics, space structures, etc.) and identify the authors according to the area they work in. Now look at the abstract or summary that appears in the beginning of the report. Glance through the report at figures, tables, and equations and then read the summary.

You are no longer trying to understand every word, you are investigating who the authors are and what they do. Pick the author who is closest to the area you're working on. Write down his name, title of his report, report number, and the date. Get out the company phone book and look up his phone number. Do not send e-mail. Do not write a note.

Take a few deep breaths to calm your nerves. Pick up the phone and call.

"Hello, this is Joe Newman, may I speak to Gary Bright?"

"Speaking."

"Hi. I'm the new guy in control analysis and I was just reading your engineering memorandum on dual-spin dynamics. Can I make an appointment to talk to you about it?"

Gary Bright will be delighted to talk to you about his work. With the exception of a very few pathological cases, all rocket scientists love to talk about the work they do. They are proud of it and they will welcome your visit.

Make appointments with all the major players you identified by reports and word of mouth. Show up on time, with a pen and notepad. Smile, introduce yourself, give a firm handshake. If you don't know how to give a firm handshake, practice with a friend. A cold fish, wimpy handshake is a turnoff to everyone—it gives a poor first impression. Don't make it a vice grip either—that gives another negative impression. Firm and friendly. Look into the rocket scientist's eyes and smile. You are glad to meet him!

"So what can I do for you?" he asks.

"I was just reading your memo and wondered if you would explain it to me."

When he speaks, listen carefully and take notes. Listen to understand. If he says anything you don't understand, stop him and ask for clarification. If he uses any acronyms, ask him what they stand for and what they mean.

Don't be afraid to ask dumb questions. You have a great excuse to ask them now. You're the new person. This is your first week on the job. Of course you don't know!

Feel free to ask away. Think before you ask. Make your questions thoughtful and to the point. If you are listening to the rocket scientist, your questions will follow naturally and intelligently. You are ignorant about Space Systems X, but you are not stupid. You are an intelligent engineer who can learn new things.

Do not overstay your welcome. When you sense that the meeting is coming to an end, rise, thank the rocket scientist for his time, shake his hand, and leave. Show your respect for him by being punctual, attentive, and grateful for his time.

Congratulate yourself: you've made a new friend and colleague.

Go back to your office. Clean up your notes. Read a little more of the rocket scientist's report. Don't try to read it all. When you get stuck, stop and write down your questions.

After you have met with several engineers you will start to understand the problem you are working on. You will begin to pick up the acronyms. Things will start to make sense. By the end of your first two weeks you will be ready to start writing your first report to your boss on the control system for Space Systems X.

Now you will have a second chance to thank all the rocket scientists who helped you. You will include them in an acknowledgment section at the end of your report. It will say, "The author gratefully acknowledges the expert advice of the following individuals..." and list them. Also, you will have a list of references in your report. You will list every author's report that you used, however slightly. Then you will distribute your report to your boss and all the authors you've acknowledged. Be sure to get the boss's distribution list as well. He knows better than you who should get your report.

Give yourself a pat on the back. You've made it through your first two weeks, you've solidified contacts with many rocket scientists and started your network, and your well-deserved paycheck has finally arrived!

Chapter 16:
Reinvent the Wheel

Even if your boss is semi-enlightened he may admonish you with the oft-repeated command, "Don't reinvent the wheel."

This has become the company creed, passed down from the highest levels of management to the engineers who must figure out how to make everything work. The company means, "Don't waste money, don't duplicate work we've already paid for, just work on the new problem we're dealing with now."

The for-profit company motives are obvious, but you should not take this decree too seriously. The company wants you, the engineer, to make things work, just as Captain Jean-Luc Picard issues the command: "Make it so!"

But for the engineers to "make it so" they must understand the system completely.

You must go back to the basics. Go to your foundations in Newton's laws, Euler's equations, Lagrange's dynamics, Hamilton's principle. Get out your textbooks from school (which you will never sell —they are your library and they keep getting more valuable to you over time). Go to the company's technical library.

Understand Space Systems X from the ground up. Start with physical and mathematical analysis. When your supervisor tells you, "Don't reinvent the wheel," ignore him. Don't argue, just ignore. How can you understand the system unless you reinvent it for yourself?

Start your analysis from the beginning. You can get away with this when you are a new employee; later, there will be less time for it as your responsibilities grow. You will be given some leeway during your six-month probationary period. Don't waste it: use it to establish a deep understanding of Space Systems X. Develop your own equations, write your own code, do your own analysis. Get your own independent understanding of the system.

Don't accept anything as true until you prove it to yourself in your own way. Don't accept shortcuts. Apply critical thinking. Take a "show me" attitude. Stand on your own two feet.

Learn from the engineers. Pick their brains, but only accept what they say if they can prove it to you. This is a critical period in your development as a rocket scientist. You must have your own mind, your own understanding. Understanding cannot be borrowed. It must be obtained painstakingly through analysis and reasoning.

Think.

Do the hard work.

You will become a respected engineer whose opinions are valued and trusted.

Richard Feynman once said that he could only understand another physicist's work if he reinvented or recreated it himself. Werner Erhard said, "The truth believed is a lie."

Do not be afraid to reinvent the wheel. It is the only way to learn about the wheel— if you've never seen one before.

Chapter 17:
What if the Rocket
Doesn't Work?

The company doesn't need a "yes" man. It needs discriminating, intelligent engineers who can identify problems, solve them, and make the system work. Or, if the system can't work, the company needs to know this ASAP to avoid "throwing good money after bad."

The MX underground missile system sounded like a good idea to everyone who worked on it. A complex of tunnels and railroad tracks would allow the U.S. to play a shell game—a sort of hide and seek—with the Russians. We could keep moving the missiles around in the underground tunnels on railroad tracks like the con artist on the street playing the shell game with three cups and a little red ball. Which cup has the red ball under it? Place your bet and lose your money!

In the MX system we were betting that the Russians wouldn't be able to figure out which missile silo the rocket was hidden under if we moved the rocket every day. The shell game would discourage the Russians from trying to knock out our retaliatory force with a first-strike, sneak attack.

Then a bright young engineer was hired to work on the MX system. He asked a dumb question, only it wasn't so dumb. He asked, "What if there's a nuclear explosion at one of the empty silos? Wouldn't the pressure wave travel unabated throughout the tunnel system and destroy the hidden missile anyway? I seem to remember learning, in one of my fluid mechanics labs, that the pressure in a tube doesn't drop off much."

He had asked a dumb question. A dumb, embarrassing question. No one had thought to ask such a question before, and he was right. The underground tunnel idea had to be scrapped. A great deal of money had already been wasted. But the young engineer prevented the wastage of billions, and, more importantly, prevented the creation of a totally ineffectual defense system.

Galileo once said that the humblest reasoning of one man is worth more than all the authorities of the world. Use your humble reasoning. Double check your answer. Triple check it. Bounce your idea off the resident guru in your section. (He's the engineer extraordinaire who has refused all offers of rising up the management chain, who loves the technical work—the recognized genius in the section.) If your idea holds up then stick to your guns.

Trust yourself and your analysis. Make the company understand. Be polite and patient. Remember that you're on a team and everyone wants Space Systems X to work.

So do you.

Chapter 18:
How to Tell Your Boss:
"We've Got a Problem"

During the Apollo 13 mission, the astronauts reported, "Houston, we've had a problem." The problem had just happened, it wasn't their fault, and they were reporting it as soon as they could.

As a rocket scientist, for the most part, you will be working on systems that are still on the drawing board, in the concept stage, in construction, in preflight testing, or in mission planning. Unless you are in launch or mission operations, you will be working months or even years ahead of the actual flight of your system.

You don't have the luxury of waiting for the problem to happen and merely reporting it. You are supposed to foresee the problem and prevent it from happening. You must let your boss know far ahead of time if your analysis indicates that the rocket system will not work as planned.

The system is expected to work, to perform according to certain specified requirements. You must determine if the system can really achieve its mission and its true capabilities.

A young man calculated the maneuver capabilities of a spacecraft. He used the laws of physics (Newton's and Euler's laws) and the hardware capabilities of the spacecraft to make his assessment. He had learned the laws of physics at school; he learned of the capabilities of the hardware by interviewing engineers and technicians who designed and built it.

The young man had been given the assignment of maneuver analyst the previous year. He had asked for the position as a condition for accepting the company's job offer. A middle-aged engineer was moved from the assignment to accommodate the new hire.

When the young man put his equations and numbers together, he found that the spacecraft would not be able to perform all of its maneuvers accurately enough. He informed his boss about the problem. His new boss was not the person who hired him. His original boss had left the company to take a lucrative job in Seattle—an offer so high he couldn't refuse it.

Unfortunately, the new boss turned out to be little more than a mouthpiece for upper management.

When the young man first informed his new boss of the problem, the new boss told him that his job was to prove that the spacecraft would work as planned. "But the spacecraft will not be able to perform balanced turns within specs," the young man said.

"Perhaps I didn't make it clear," the boss said. "Your job is to show that the spacecraft meets all the requirements."

The young man returned to his office, perplexed at his dilemma. He checked his analysis again. He went over the tolerances reported by the technicians on the accelerometers, gyros, thruster offsets, structural misalignments, and mass property uncertainties. His work checked out. He made another appointment to see his boss.

"The last maneuver analyst and the analyst before him reported that the spacecraft will perform as planned," said his boss. "Why should I believe you?"

The young man replied, "I have been over their calculations. They used back-of-the envelope methods, which can sometimes be good, but not always. I have used an in-depth analysis. I've documented all my work in three engineering memos totaling 250 pages. If I am incorrect, show me where I made an error in my equations or in the numbers I have quoted in my references."

The boss was impressed, but doubt plainly showed on his face.

The young man added, "These are the results if you take physics and apply the numbers from the hardware."

The boss decided to withhold his judgement. The young man stuck to his guns. Two years later the navigation team accepted the young man's analysis: balanced turns were inaccurate. They would have to rely on unbalanced turns for most spacecraft turns. The young man first gained the respect of other rocket scientists. He was polite and patient. Eventually he also gained the respect of the management. His reputation and responsibilities grew. "All I want is to make sure the spacecraft will work," said the young man.

The young man was a team player, but he had to play an important singular role as the maneuver analyst. In your job as an aerospace engineer you will be called upon to make an honest and accurate assessment of a rocket system or space vehicle. Make sure you do your analysis carefully. Make sure you are right. Then, stick to your guns.

Do not be intimidated by the desires of management. Rather, be motivated by the success of the project. Be polite, patient, and truthful. Document your work. Give presentations. Be open to criticism. If an error of yours is pointed out, be grateful.

You are on a team. You want the team to be successful. It is much like a baseball team. When the ball comes to you, you're on your own—you have to perform your function well and no one else can do that for you. The winning team consists of several individuals all performing their functions well and in concert with each other.

You must do your part to help your company win. Don't focus on the narrow limits of your job description. Take responsibility for the success of your company's mission.

Then you will not have to hear later that "we've had a problem."

Chapter 19:
Keep Your Boss Informed

Keep your boss informed about your progress.

You must be absolutely clear about what your responsibilities are. At the beginning of each new task, meet with your boss to establish two things:

1) What products you are to deliver; and
2) When they must be delivered.

The products may include an engineering memorandum or report, computer code, or a presentation. They may involve analysis, mathematical modeling, simulations, or laboratory tests.

It is paramount that you understand exactly what your boss expects of you. This is the time to ask the dumb questions. Should you write a lengthy report? How long? Will you be making a presentation? Who will the audience be? How long will you speak? Should you wear a suit? (More on report writing and presentations later.)

The delivery date is just as important. There may be several dates when you turn in your results in stages or installments. Above all, you must not turn in any product late. No excuses are acceptable. Maybe your professor accepted late papers for any number of reasons. "The copy machine broke down." "My Great-Aunt Emma died." "There was a lightning storm and my hard drive crashed during an electrical surge." "The cat ate it."

Put your excuses away. You are in the big leagues now. You are getting paid good money to deliver good product on time. If the rocket lifts off at 12:01 p.m. Eastern Standard Time, delivering the launch guidance code at 12:02 p.m. will not do.

A late report is a breach of trust between you and your employer. It will not be forgiven. It will affect your raise, your promotion, and perhaps even your continued employment.

When you set up your schedule with your boss it may not be clear even to him if all the tasks you are assigned can be achieved in the allotted time. There may be unforeseen difficulties. The plan may be unsound. Some new device or technology may be required that does not yet exist and delays progress.

If you unearth such difficulties, do not hide them from your boss. Let him know at the earliest possible moment.

Do not drop a bomb on your supervisor with a final report entitled "Why Space Systems X Is Destined to Crash on its Maiden Flight." You must tell your boss bad news even sooner than good news. (And be sure you are right about it.) You must do this very early—well ahead of any scheduled product deliveries.

Explain the problem to your boss. First give him the bottom line. If Space Systems X cannot fly in its present configuration, tell him this in the privacy of his own office, behind closed doors. Be prepared to prove the results of your analysis step by step. Also be prepared to offer alternative plans toward a solution. (This may be difficult.)

Your boss is intelligent—he will understand the problem and will bring new resources to bear on it, if warranted. When your boss recognizes the difficulties, you will be in a position to renegotiate your product delivery dates. New tasks will be assigned to you with a new schedule.

Your boss may bring in help. He may inform upper management. The problem must be solved. You must always maintain the right attitude: the problem must be solved.

As Gene Kranz said, "Failure is not an option."

Chapter 20:
Reality Therapy: A Few Words About the Challenger

The engineers were against it. They didn't want to launch *Challenger* on that cold January day in 1986.

Icicles were hanging off the engines. The engineers had no flight experience at low temperatures. They were aware of scorch leaks through the boosters' O-rings on previous missions, and they suspected that the leaks were correlated to low ambient temperatures.

But management wanted to launch. They had to prove that the Shuttle could fly frequently and safely to demonstrate its cost effectiveness. Also, the president would be giving his State of the Union address and (there were rumors circulating within NASA that) there was to be an exchange between the president and the Shuttle crew, including the first teacher in space.

NASA had sold the Shuttle to Congress by stating that it would fly regularly like an airliner and that its probability of failure was only 1 in 100,000. Richard Feynman, the Nobel prize-winning physicist, commented that this would mean that we could fly every day for 300 years before it crashed. It defied common sense. The rocket scientists knew that the best success record of any existing rocket system was 99%: a failure rate of 1 in 100. Many rockets crash far more often, particularly in early flights.

On that fateful day in 1986 one of the managers talked to the top technical manager, an engineer who managed launch operations: "I want you to take off your engineer's hat and put your manager's hat back on and make a decision. When do you want to fly again, next April?"

The correct answer (as we all know now) was, "Yes I'd like to be flying next April, but if we crash in January we won't be flying again for another two years!"

The technical manager put on his manager's hat and approved the launch. He buckled under the pressure. The *Challenger* exploded 73 seconds after the launch, killing seven astronauts and putting the space program on hold for two years.

We see in this example, this national tragedy, the two powerful forces that control high-profile space systems: politics and engineering. It is as if there are two realities existing side by side: political reality and physical reality.

The politicians are well represented in the film version of *The Right Stuff* where a manager smugly asks a scientist, "Do you know what makes those rockets go up?" When the scientist starts to explain the complications involved he is interrupted with the answer, "Dollars make those rockets go up. No bucks, no Buck Rogers!"

And it is true, but only partly.

Richard Feynman eloquently rebutted the lopsided view of the politician when he said, "For a successful technology, reality must take precedence over public relations, for Nature cannot be fooled."

It is your responsibility to understand the science and engineering of the rocket system. You will sometimes be asked to make things work, to fudge numbers, to say everything is A-OK when your analysis says it is not. You must have courage. Stick to your guns. Do not let your company do a dumb thing. Patiently explain the problem to your managers.

As a rocket scientist, you know that you cannot fool Mother Nature. You may receive orders from the top of your company to do otherwise. A great deal of money may be involved. The president may be giving an important speech. But you know better than to confuse political reality with physical reality.

Some day you will be tested. It may be on a minor matter or it may be of great significance. That is when you will need all the courage you can muster. That is when you and others will learn about your character.

You will not be alone. You will have Galileo and Newton and Euler and Lagrange on your side. You will have your fellow engineers behind you.

Do not be afraid to make the right decision and to stand by it.

Chapter 21:
Work on the Big Picture

Two bricklayers were busy laying bricks. A young boy came by and asked the first bricklayer what he was doing. The bricklayer looked disturbed at the interruption. "Why, can't you see? I'm laying bricks!"

The boy walked over to where the second bricklayer was working and asked him the same question. The second bricklayer stood up, smiled and motioned with his trowel over the construction area. His face lit up and he said, "I am building a cathedral!"

The first bricklayer was working only on his immediate task. His job description. He didn't care about the overall project.

He was a little-picture person.

"I just do what they tell me," he would say. He derived no great pleasure in his work.

The second bricklayer was a big-picture person. He derived meaning in the task. He always said, "I'm glad to be a part of it." It made him happy to envision the beautiful cathedral he was helping to create.

The second bricklayer was also laying bricks. He was working on the little picture and the big picture.

The lesson is obvious. If you want to be a great rocket scientist, you must work on the big picture. Pay attention to the overall success of the project. Do not let a task fall through the cracks because it isn't in your job description.

A young engineer informed his supervisor that the spacecraft would not be able to control a trajectory correction maneuver accurately enough to meet the specifications.

"Why not?" the boss asked.

"Because the gyros will not operate in all-spin mode and so we won't have updates on the current spin rate," said the young engineer.

"So?"

"So the accelerometers will not get accurate spin rate and therefore will not be able to control the burn."

"But the control analysts have promised us that they will provide current spin rate," replied the boss. "It's been documented in the specs, so it's not our problem."

The boss started to search for the document.

The young engineer shook his head. "It doesn't matter what they promised," he said. "They can't supply the needed information because the spacecraft doesn't work that way."

"So what do you want to do?" asked the boss.

"I want to make sure the spacecraft will work. We can break up the burn into two stages: a 90% burn and a 10% clean-up burn. If the first burn is 10% high, we don't perform the second. Otherwise, we correct with the second burn. This will make up for the errors from not having current spin rate."

The boss was working on the little picture, but he quickly saw the young engineer's point and agreed with him. The young engineer was working on the big picture. He gained a reputation as a problem solver. He identified problems early and laid out practical solutions to them.

The young engineer was happy to receive an excellent raise that year. But he was prouder of the respect he gained from his colleagues and his boss.

Chapter 22:
How to Give a Presentation to Rocket Scientists

You have done the hard work (as described earlier): you have performed a careful analysis and now you have something important to say. How should you present your work to rocket scientists?

Remember that rocket scientists are human beings. When you give a presentation to rocket scientists you should realize that they can get bored as easily as anyone else. You can make their eyes glaze over with too many equations, too many words, too many slides. You can put them to sleep as easily as any other human being if you subject them to a dull lecture or if you talk too long.

So the ordinary rules of how to give a good presentation still apply, with minor exceptions. (See, for example, Reid Buckley's *Strictly Speaking*—it's excellent!) The exceptions involve the technical background of your audience.

You can and sometimes should have equations, but keep them to a minimum. You should have some slides, that is, you shouldn't eliminate all slides but also keep them to a minimum.

The golden rule of presentations is brevity. Keep it short!

Reid Buckley recommends no more than 30 minutes for any lecture. This is a very good rule that you should strive to follow. The rocket scientist may violate the 30-minute rule in some special cases, but brevity is still golden.

Brevity can be thought of in terms of expectation. If the audience expects a one-hour lecture, then make it shorter: no more than 40 minutes. Use a two-thirds rule for your absolute maximum. If you are given 30 minutes, don't talk for more than 20 minutes. If you have 15 minutes, finish in 10.

Brevity is a matter of perception: your talk must appear brief in terms of expectations. You must gauge your audience.

If you are giving a one-hour seminar on, for example, "Mission Design: Theory and Practice" to mission designers at Caltech's Jet Propulsion Laboratory, then you should talk for only 30 minutes, or half the allotted time.

Why?

Because mission designers at JPL are extraordinarily inquisitive and intelligent. They will interrupt you to clarify a point. (They will clarify your point for you.) They will add their suggestions to your seminar. They will start a discussion in the middle of your lecture, creatively jumping off your starting point and elaborating on it and exploring it.

They will play with your ideas, laugh, tell jokes. They are like children—they want to have fun.

Let them have their fun. If you talk for 30 minutes and they talk for 30 minutes, they will conclude that you are one of the most stimulating and interesting lecturers they have ever listened to!

On the other hand, if you are given one hour to present your seminar at an academic institution in pursuit of an assistant professorship, you should probably talk for 40 to 45 minutes.

Why?

Because you will get much less interaction from a room full of professors and graduate students than from a room full of JPL mission designers. Professors do not like to reveal their ignorance. They are custodians of knowledge and are assumed to know everything. This is a difficult burden to carry and they learn defense mechanisms to protect themselves from exposure. They are not freewheelers.

They will, perhaps, ask a stodgy academic question. There will be one expert whose job it is to attack you and demonstrate your inadequacy for an academic position in their lofty institution.

The others will know little of your work and will be totally harmless.

This audience expects you to fill the hour. If you finish early they will conclude that you have run out of material, that you are not deep.

It is very difficult to prepare such a seminar and there are many pitfalls. (More on this later.)

If you are a graduate student preparing a thesis defense, your presentation will be scheduled for about two hours. Do not assume that you should be talking that long! Plan for 30 to 35 minutes, no more. Your faculty committee members are busy people. They don't want to listen to you for long.

"But I've been working on my thesis for three years! I want to show what I've done. Won't they think I goofed off all that time if my defense is only 30 minutes long?"

No, because you gave your committee copies of your Master's or Ph.D. dissertation two weeks before your defense. It is a substantial document jam-packed with theoretical analysis, mathematical models, equations, simulation results, tables of numerical data, figures, and plots. Also, your committee knows your advisor. They will vote with your advisor to pass you, provided you don't bore them to tears with a long-winded, two-hour presentation!

If you are working on a conference presentation, you may have only 15 minutes. Within that 15 minutes you have to be introduced, present your talk, and answer questions. You must allow five minutes for the Q&A period and one minute for the introduction. Expect a one-minute delay before the session chairman introduces you. In this case, prepare a talk that is eight minutes long.

Remember that there may be hundreds of papers being presented at a conference. Usually there are several sessions going on simultaneously, so the talks have to stick to a tight schedule to allow the audience to move from one session to another and to hear the talks they are most interested in.

"So what if there is no delay and my introduction takes 15 seconds and I only get one question? Won't I look like a fool sitting down after only nine minutes?"

Not at all. The people in the audience will be grateful for a break in the lectures. They can finally talk. Some will come up to you to ask questions they were afraid to ask in front of the audience. Some will have to relieve their bladders. Some will need to run out of the building to smoke a cigarette.

Others may be looking to hire a bright rocket scientist like yourself. Have your business card ready. It's not a bad idea to have your resume ready also.

If the session chair is behind schedule, he will be grateful to you for giving him back six minutes.

Everyone is happy when a lecture ends early!

Chapter 23:
How to Keep Your
Presentations Short
and Snappy

Y ou keep your presentation short and snappy by keeping your slides short and snappy.

You should use no more than two slides for every three minutes of your talk. That is, plan to talk an average of one and a half minutes per slide. Count all of your slides including the cover with the title of your talk and your name.

It is crucial that you do not exceed the slide count. Do not argue, "I can talk faster" or, "This slide only takes a few seconds" to stuff a few more slides into your presentation.

Do not cram more information into each slide to meet the slide count. Each slide should be a model of brevity. Simple, concise, to the point. In a word: snappy.

Avoid extraneous words, equations, graphs, and tables. Avoid extraneous information of any kind.

Your audience can only absorb so much information during the allotted time. You must edit and edit severely.

Use large font. Don't make your audience squint—you aren't designing an eye chart. Use the floor test. Throw your slides on the floor. If you can read them easily from a standing position, you've past the test.

When you start putting your slides together you will usually be working from a document you wrote: a technical report, a dissertation, or a conference paper. Begin with this document. Select figures, plots, and crucial equations from your document. Go for the highlights, not the details. Avoid repeating similar material.

No figure should look alike. No analysis should follow the same path. If you proved a theorem in discrete time, do not provide the parallel proof in continuous time, merely state that you have done it. If you have a plot showing oscillations with light damping, do not provide a second plot showing oscillations with slightly less damping, unless this is your crucial point.

After you have selected materials from your document, put the document aside. If you have no document to start from, then what follows is where you begin to formulate your presentation.

You are about to create your critical slide. It is Slide 2. (Slide 1 is your title and name.)

Slide 2 should be entitled: "The Problem." It can be a sketch or photograph. This slide should epitomize the problem you are working on. It could be a picture of your rocket lifting off, your spacecraft encountering Jupiter's moons, or your lander descending to Mars.

You should talk for about three minutes on this slide. Give a short history of how the problem arose, why it is important, or why you find it interesting. When you are done talking about Slide 2, no one in the audience should have any doubt about what problem you are trying to solve.

Now they are interested because they understand your goal. You have created drama because you have not told them if you achieved it or how you achieved it. The audience is now on your side. They are rooting for you. They are eager to learn how you dealt with such an important and difficult challenge.

On the other hand, if you have failed to firmly plant your problem in the audience's mind, then you have lost them. Permanently. Forever.

They're not coming back.

Nothing you say later will bring them back.

And it's all your fault.

Furthermore, if you have lost your audience in the beginning, you only frustrate or bore them the more you talk. So if Slide 2 fails, you will have either

1) an angry audience, or, less threatening,

2) a drowsy, falling asleep, even snoring audience.

"But I have so many important things to tell them. Why should I waste my time on a stupid cartoon, just talking about the problem—not solving it?"

Because if you don't bother with the "cartoon" nobody will care what "important problem" you might have solved. People don't care about things they don't understand. You must go back to the beginning and make them understand.

"What do I have to do—go back to kindergarten?"

By all means, yes! Go back to basics, go back to the beginning.

Even rocket scientists like to be reminded of the foundations of physics and mathematics upon which all their knowledge is based. It's like telling a fairy tale to children. They like to hear it over and over again. It's OK to mention Newton's and Euler's laws. Even desirable. People are comfortable starting with the familiar. Say things everybody knows.

Slide 2 is no joke, however. You're not goofing around. Slide 2 is serious business. (But you are allowed to joke about it.) It is your business. You should love the fundamentals of your business.

In your next slides, you will tell the story of how you dealt with the problem.

Do not kill your momentum by putting up a slide that (essentially) says:

OVERVIEW

Beginning

Middle

End

Do not say, "This is my overview slide. First, I will give an introduction—which I am doing now—then there will be a middle, and, finally there will be an end, when I conclude."

(Do this only if you want to appear like the moron, Chance the gardener, in the movie *Being There*.)

You have set the stage with Slide 2. Don't outline your story. Would you like to read a mystery novel that gives you an overview on page 3?

Jump into your story on Slide 3. You may not have much time. If your talk is to be five minutes, then this is your last slide! Now that some companies are requiring their employees to learn to give one-minute presentations, a five-minute talk is certainly not out of the question.

If Slide 3 is your last slide, you will cut to the chase and give the bottom line of how you solved your problem. Clearly you will have to skip the details.

But even if your talk is 30 minutes, you should have a maximum of only 20 slides. Even if you are presenting a Ph.D. dissertation on analytic theory of ballistic entry into planetary atmospheres, you should not have more than 20 slides.

You should favor pictures above all else in your slides. Why do we say a picture is worth a thousand words? Because we are visual creatures.

Use words or phrases only when pictures cannot be made to convey your ideas. When you resort to words, avoid complete sentences on your slides. You will speak in complete sentences, but you should not write them out on your slides. The exception is when you have an extraordinarily important statement or conclusion that carries its greatest impact as a sentence.

Avoid equations like the plague.

"But I am an analyst. I live for equations. My equations are my life's work!"

Throw them overboard.

You are talking to human beings (albeit rocket scientists). You must convey your ideas with pictures and words.

Above all, do not give derivations. Do not give mathematical proof of theorems. You can give the gist of the theorem, a sketch of the proof. Use a picture to illustrate it, if you can. Do your utmost to create pictures of your ideas. Draw on your creativity.

This is important.

Do not despise the process of making your ideas clear. Do not make a half effort—put all your ingenuity into it!

You will become a master of presenting complex ideas. You will be famous. You will be well paid.

All right, you may have a few equations. One or two equations on one or two slides. If your Ph.D. dissertation is largely mathematical, you can have more equations, but keep them concise. Put your equations in highly simplified form:

$$y = f(x)$$

You can tell your audience "the function f contains nearly 100 terms consisting of Bessel functions, Fresnel integrals, and elementary functions."

Don't write out the 100 terms on your slide! You can say, "The details are provided on pages 92–94 of the thesis." Then move on to your numerical results. You can make a plot of your equation $y = f(x)$ with y on the vertical axis and x on the horizontal.

When you describe a plot, you should first identify what you are plotting. "Here I am plotting the function f versus x." Next, point out the axes and specify their ranges. Use your pointer precisely. Say "The x axis ranges from 0 to 1000 km," and point at the zero and trace the x axis to 1000 km. Then say, "The y axis ranges from 3400 to 3500 km," and then trace the y axis from 3400 to 3500 km.

This is a courtesy you are giving your audience. Rocket scientists will immediately look for the axes and the units. It takes time to see the axes, to understand what is being plotted against what, and to reflect on the units, magnitudes, and meaning of the plot. You already know what's on the plot. It's trivial to you. You've seen it a hundred times. Your audience is seeing it for the first time. Be courteous. Give their eyes a chance to adjust. Point precisely to where their eyes should focus.

Trace the function, $f(x)$, precisely and say, "We see that the function, f, rises from a minimum of 3410 km, to a maximum of 3490 km, then descends rapidly to 3450 km."

Then put the pointer away. Do not point it at the audience (even if it's not a laser pointer). Do not shake your pointer at the audience. That is impolite. Do not wave it ambiguously at the plot. That causes confusion.

Use the pointer precisely then put it away. You can hold it alongside your leg, out of sight, until you need it again.

If you have been careful in describing your first plot, the audience will be with you and you can move more quickly on subsequent plots.

You should have only one plot per slide. You may have more than one function (contour) plotted, but avoid having more than a few. Dozens of functions plotted on one plot are confusing and useless.

Edit, simplify, reduce to the essence. If your audience understands one function, they will understand you if you merely state, "I plot dozens of related functions in the paper."

If you have clearly conveyed your problem, your analysis, and your results, then your audience will be ready for your conclusion. In your conclusion, do not tell your audience what you just did. They know what you did. You communicated your ideas clearly and they understood.

In your conclusion, put your results into perspective: what does it all mean? You did all this work and you solved your problem—so what?

The conclusion gives you an opportunity to summarize not what you did, but why it is significant.

"The new launch guidance system passes all tests with flying colors—now we can look forward to safe, successful launches in a wide variety of wind conditions. Thank you for your attention."

Congratulations! Your audience will remember what problem you solved and why it is important. They know what you did and why you did it.

Chapter 24:
How to Write
a Technical Report

When you write a technical report or memorandum, you are writing for three audiences:

    1) your boss (the management),

    2) your colleagues (engineers), and

   3) yourself.

Keep this in mind as you write the different sections of your report. (Also note that where the following suggestions differ from your company's format, you should adopt your company's structure and style.)

The first page of your memorandum should provide a summary. In the summary, you compress the whole story of your report into approximately 300 words or less. The summary is equivalent to the abstract of a journal publication. In very few words you give the origin, significance, and solution of the problem.

The summary is meant to be read by everyone, but it will be particularly important to your boss and to the management. If management reads your memo at all, they will read the summary, flip through your analysis, glance at a few figures, and concentrate on your conclusions.

Your conclusions should be directed at your managers. If you are reporting bad news, your boss and his bosses will be scrutinizing your conclusions.

Write your conclusions in a series of numbered statements that are pertinent to the success of the mission. Do not repeat what you have written in the summary. Do not repeat whole sentences from the introduction. Do not subscribe to the philosophy that says:

    1) first tell 'em what you're gonna tell 'em,

    2) then tell 'em,

    3) then tell 'em what ya told 'em.

This kind of writing is boringly repetitive. It is insulting to the reader.

Your boss will read your summary, then your conclusions. The conclusions should be a continuation, not a repetition of the summary.

Your summary may say something like: "The spacecraft maneuver capabilities are addressed by analytical and numerical techniques. All maneuvers except the balanced turns meet specifications. Recommendations for dealing with balanced turns are suggested."

Then, in your conclusions, you can say the following:

"1) Axial maneuvers meet pointing and magnitude specifications provided the maneuvers are greater than 0.1 m/s and less than 10 m/s. Large maneuvers must be broken up into a sequence of maneuvers less than 10 m/s each. Between each axial maneuver a spin-rate-correction maneuver must be performed to prevent spin rate from exceeding specs. The axial torques from thruster misalignment and jet damping are responsible for this phenomenon.

2) All other maneuvers perform as planned except for the balanced turn.

3) Balanced turns cannot achieve maneuver requirements because of the large mismatch uncertainty in the thruster couple (10%). Recommendations for handling this problem follow.

4) Unbalanced turns can be used in conjunction with axial and lateral maneuvers instead of balanced turns. The unbalanced turns are more accurate than the balanced because only one thruster is used. (The error is 5%.)

5) Balanced turns may be calibrated during flight so that the uncertainties could be reduced. If this strategy is adopted, further analysis will be necessary to confirm its predictability."

When your boss sees your report it should not be the first time he learns of this problem. He must be informed of it by you far ahead of time. And your boss must agree with your strategies to fix the problem before you write the report.

After the conclusion section, you should have an acknowledgments section. This is where you thank the other rocket scientists whose advice was so important to your analysis. After the acknowledgments is the reference list where you list all the important company memos and reports you have used, however slightly. You will probably have a few textbooks in your references as well.

Your acknowledgments and references show how you value the support of your fellow engineers. It shows your appreciation. It wins friends and influences people. When your memo is distributed, the people you acknowledged or referenced can point out to their bosses how they helped the new employee. Bringing new people up on the learning curve is important to the success of the company; those who helped you get a good start will also be rewarded.

And everyone loves to see his name in print. Your colleagues will be happy to work with you again when they see that you have thanked them in black and white.

The acknowledgments and references establish your network within the company. Your network extends your ability to make valuable contributions. As your network expands, you become a more visible and valued employee. (More about visibility later.)

Your introduction immediately follows your summary. It should discuss the origin of the problem, refer to previous work done by yourself or others (here's your chance to start referencing your colleagues), and present your new approach to solving the problem or to take it to the next step. Do not repeat whole sentences from your summary! Your reader has just read your summary and will become bored, insulted, or annoyed when you say exactly the same thing twice. Your reader will lose interest. Avoid this common mistake.

The introduction should not go any further than to mention how your approach is different. It should not state what the results are or what your conclusions are. Move on to the body of your report.

The body of your report contains your analysis: theory, modeling, analytical solutions, numerical results, tables, and figures. It can be highly technical, but should still be written in a clear, logical fashion. You write the body of your report primarily for yourself and your immediate colleagues (for example, the maneuver-reconstruction analyst and the attitude-control flight-software analyst).

In the body of the report, feel free to put all the details you need to remind yourself how you solved the problem. You will be asked a year later what you did—be sure you can read your own report and understand it!

Use a clear, simple writing style. Avoid the passive voice: "In this paper the following problem was solved." Use active voice in the present tense: "We analyze the maneuvers of the Space Systems X spacecraft." Do not be afraid of the royal plural (we). When you avoid using "we" or even "I" you are forced to construct passive sentences. (There are editors of technical journals who still don't realize this, heeding instead the scoldings of long-dead schoolmarms who warned of the dangers of the informal, and therefore incorrect, "we.")

If you keep in the present tense and use the active voice, your writing will be more alive, more interesting, and easier to read.

If you stick to past tense and passive voice, your writing will be tired, boring, and difficult to read.

If you refer to a book or journal article, keep it present in the active voice (Greenwood discusses Lagrange's equations in Chapter 6) instead of past tense, passive voice (Lagrange's equations were discussed by Greenwood in Chapter 6).

A few passive, past-tense sentences may not seem to be any worse than active present, but when you string dozens of these weak sentences together the total effect can be positively dreary—an excellent prescription for insomnia!

When you write the body of your report, keep in mind the guidelines for giving good presentations. Hit the high points early. Take a big-picture approach from the beginning. Make clear illustrations. Choose titles and subtitles that clarify the big picture. Keep the organization in a top-down structure; that is, move from big ideas (big picture) down through highly technical, detailed ideas (little picture).

Always strive for clarity. Choose variable names that are mnemonic, that is, easy to associate with their meaning ("a" for acceleration, "F" for force, etc.). Make well-illustrated diagrams. Break complex illustrations into several separate pictures—from overall to detailed. Break everything into easily digestible steps.

Never throw out the simple picture (or idea) that you started with. If you began thinking that the velocity distribution is spherical and later (after much analysis) found that it is better modeled as a triaxial ellipsoid, don't throw out the simple spherical picture!

Keep your initial, simple approach. Keep the Zen mind, beginner's mind. Don't start your report with, "The velocity distribution of an arbitrarily misdirected injection maneuver is a triaxial ellipsoid with the major axis aligned with the initial orbital velocity and the minor axis oriented along the radial position vector from the center of the Earth to the booster."

This statement may be highly precise, but it is hardly incipiently clear. Instead say, "The velocity distribution can be approximated by a sphere. Deviation from a spherical distribution will be discussed later."

Here we note that "clarity" and "accuracy" are antagonistic.

Unfortunately, most rocket scientists are a little anal retentive and cannot bring themselves to write a statement that isn't exactly correct or, at least, highly accurate. As a result, they start with a highly complex, precise statement, replete with adjectives and modifying phrases to be absolutely correct.

They lose their audience immediately! (Even their closest colleagues will have trouble reading their report.) They have thrown clarity overboard in favor of accuracy, without the slightest sympathy for their readers!

"I wrote a correct, accurate statement. If the reader isn't smart enough to figure it out, he shouldn't be reading my memos!"

If that's your attitude, then no one will be reading your memos and your work will drop out of sight. You will lose visibility. You will become invisible.

If your writing is hard to understand, it's your fault, not your reader's. This is the cardinal rule of writing, but you were not taught this in engineering school.

If this is your problem, then you must let go of your stubborn attitude of being right and become concerned with the problem of being understood. Remember how you started your analysis: you thought of the velocity distribution as a sphere—it was easy to picture and you progressed quickly. Later, you found that the distribution was actually a triaxial ellipsoid—less easy to picture, but nevertheless true.

Start the body of your memo with the simple picture: the spherical distribution. It is clear, but not exactly accurate. But clarity is more important in the beginning. Clarity is crucial for communication.

Remember when you started your job and couldn't understand a single memo? You found that reading technical reports is extremely difficult. This is not because the subject matter is so difficult to understand, though many rocket scientists would like to believe this. When you talk to the rocket scientist who wrote the report, you find that it is really quite easy to understand!

It was his *writing* that was difficult!

Work hard on communicating your ideas. Worry about accuracy further into your report. Strive to make your writing as easy to understand as possible. Take the work-load off your reader and put it on your own shoulders.

Kurt Vonnegut said that the reader has the tremendously difficult task of translating a bunch of black dots into pictures. Don't take the attitude that your reader is a fool. Your disdain for your reader will show.

Don't be afraid to be simple in your writing, to start with the basics, to tell an interesting story. You are participating in a great adventure (as it goes in the introduction to the original *Outer Limits*); make your readers understand you and let them feel your enthusiasm.

It will not be easy. Easy writing makes for difficult reading. Easy reading means that the writer has worked very hard.

Does this mean you have to be simple minded and stick to the easy-to-understand topics? Of course not! You will develop the skills to express the most complex of ideas by the easiest series of small steps. Your colleagues will appreciate it. You will thank yourself when you read your own memo a year later. Your bosses will start reading the body of your technical reports.

Your ability to communicate well in your writing will make you a highly valued employee among rocket scientists. Your visibility, responsibilities, and salary will rise accordingly.

Chapter 25:
The Importance
of Being Visible

You can do a wonderful job on your assigned task—accurate analysis, impressive presentations, and excellent engineering memorandums—but if you don't have visibility it will profit you very little.

Yes, you will get above-average raises and a pat on the back from your supervisor, but you will not rise within your company to the top salary unless you achieve visibility. (If you are not that ambitious, then you need not read further. There is nothing wrong with setting modest goals and enjoying your life.)

Visibility means that upper management is watching you carefully because you are working on a crucial task. The fate of Space Systems X depends on your correct analysis or ingenious design. The future of the company itself is at stake.

Visibility means that you feel the pressure of the company—you are on the "hot seat." You will know when you have visibility because you will carry great responsibilities.

Do not fear the difficult task; you can meet the challenge. It is important to remember that you only have to do an adequate job when you are highly visible. You only have to be successful. You don't have to be extraordinarily successful.

When Captain Jean-Luc Picard orders the helm to "make it so" that's all he expects. Nothing fancy. He's looking for reliability and punctuality.

If you have a highly visible task, you must be reliable and punctual: you must turn in adequate results on time.

No excuses.

Most young rocket scientists make the mistake of thinking that if they work hard it will automatically pay off. If they come to work early and work late, it will be appreciated. If they do a perfect job, they will be rewarded. They will be promoted and their salaries will be increased dramatically.

Doing an extraordinary job on an unimportant task will get you nowhere. Your boss isn't going to promote you above himself. In *The Peter Principle*, Peter and Hull warn about trying to "push" your way up. What you want instead is "pull": the attention of upper management. Having pull is equivalent to being visible. According to *The Peter Principle*, it is better to pull than to push.

It is not easy to achieve visibility, but if you are aware of the principle you already have an advantage over your fellow rocket scientists.

Chapter 26:
How to Achieve Visibility

Do good work: be reliable, accurate, and on time. Create a network, acknowledge others, and distribute your work widely.

Work outside the box. Besides doing your own job well, learn what others are doing. Don't stay within your job description. Be curious about everything relating to your project. Attend presentations and meetings outside your area. Find out what's going on across the board.

Be ready, willing, and able to take on new tasks and responsibilities. Volunteer for more work. Learn new techniques, new analysis, and new software. Take technical courses the company offers or take graduate courses at a nearby university. Your company will pay for your additional education and then reward you with a raise. Earn an advanced degree.

Perform research work. Take your specific tasks to a greater level of generality. Ask, "What is the best way to solve this problem?" "How can I solve a cluster of related problems?" "What is the most creative answer?" "What is the theory behind the specific problem?"

Publish your work. Present it at one of the AIAA conferences. Publish it in the journals. You will establish a name for yourself across the country and the world.

Become an inventor. Apply for provisional patent protection. Consult with your company's technology transfer office. The company will pay for your patent application, and you may receive a percentage of the royalties. Some companies make you sign over these rights when you become an employee, in which case you will receive little to no royalties. But even if you get just a plaque and a certificate, you do achieve visibility within the company as an inventor.

Identify the key tasks of the project. Make lateral moves to acquire skills in these areas. (A lateral move is a move to another job within the company with no salary increase or promotion.) You cannot expect to be promoted to the top of your company without understanding nearly everything about your company. When Bill O'Neil, the project manager of the Galileo mission to Jupiter, was asked what it took to rise to the top of one of NASA's most complex space missions, he said: "You've got to be interested in everything!"

If you are not given new responsibilities, or promoted rapidly enough, take a new position in another section of the company—in one of the key areas. You should make a lateral move every three years or so if you aren't promoted. Three years is enough time to learn 80% of the new area. You would have to spend another decade to learn the remaining 20%. Now consider the difference between two employees: one who spends nine years perfecting his skills on one task, and a second who spends three years in guidance and control, three years in navigation, and three years in mission operations. The first employee will be a respected expert in his area. He will have honed his skills as a specialist. He will also have reached a salary plateau. He is a necessary component of the company and is well paid. But he is also unpromotable because he is too narrow. He is, in fact, afraid of change and has become comfortable in his role as an expert.

If you allow yourself to get comfortable in a task, you may be in for a shock when the company no longer gives you excellent raises, when you are not offered the promotion to supervisor, and when you can no longer make a lateral move either within the company or out of it.

Many rocket scientists go through a slump period approximately 10 years into their careers. It is the seven-year itch of rocket scientists. They are no longer satisfied with their jobs and are helpless to change the situation. They have not planned for this.

They did what they were told.

In the movie *Falling Down* the hero (magnificently played by Michael Douglas) did what he was told and ended up dead. His last words were, "I'm the bad guy? How did that happen? I did everything they told me to." He was truly shocked.

You can avoid the shock of finding yourself underappreciated by taking an active role in your career. Do not leave it up to your supervisor. You must plan. You must make changes. You must be willing to take risks, change jobs, take courses, and learn new skills.

You will have to work with new people. Change your circle of friends. Feel stupid asking dumb questions all over again.

Don't be afraid. Your old friends and colleagues will still be yours. You will find new friends and new colleagues as well.

So make a change every three years. It's like picking up another Master's degree. Acquire skills that make you indispensable to the company. Develop a network all across the company. You will be well known and respected by your colleagues and you will obtain the attention of upper management. If you publish journal articles or present conference papers, you will become known around the world.

Keep track of your accomplishments by recording them in your resume. Always look for ways to add to your resume. If you follow this plan, then in nine years you will have worked in three areas of the company, taken graduate courses or degrees, presented conference papers, and published in archival journals.

Keep your resume and business card up to date. You may decide to move to another company. Be ready for opportunities elsewhere.

Your strong resume is not only a ticket to greater opportunities but is also insurance against company politics. A manager may decide that you are no longer valuable to the company, but your resume is strong evidence to the contrary. You cannot expect to please everyone. Please and protect yourself. Invest heavily in your resume.

You are a valuable person. This may not be recognized in the heat of local politics, but on the global scale it will. Your strong resume is your ticket out of a bad situation and into a great opportunity. Take care of yourself: take care of your resume.

$$\overset{i}{\vec{a}}{}^{P} = \dot{V}\hat{v} + \overset{i}{\vec{\omega}}{}^{4} \times V\hat{v}$$

Part V

How to Land an Academic Position

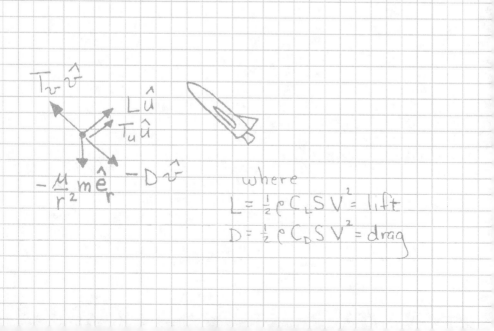

$T_{v}\hat{v}$

$L\hat{u}$

$T_{u}\hat{u}$

$-\dfrac{\mu}{r^{2}}m\hat{e}_{r}$ $-D\hat{v}$

where
$L = \frac{1}{2}\rho C_{L} S V^{2} = \text{lift}$
$D = \frac{1}{2}\rho C_{D} S V^{2} = \text{drag}$

Chapter 27:
So You Want to be a
Professor of Rocket Science

Much of the preceding advice also applies to those who want to pursue a faculty position in rocket science.

There are, of course, additional requirements you must satisfy to be considered for such a position. In a nutshell, these are as follows:

1) You must have a Ph.D. in aerospace engineering (or aeronautics and astronautics) or in a closely related field from a prestigious university.

2) You must have an established publication record in academic journals.

3) You must have funding (or credible funding contacts) to support your research work.

These three items are listed in increasing order of difficulty. Even so, the first step of getting a Ph.D. is quite a challenge.

Incidentally, while this advice traditionally applies to those who seek a position at a research university, there is a trend even among teaching institutions to require all three items.

Chapter 28: Qualifying for the Ph.D. Program

"**I** think I'll go back to the university and pick up my Ph.D.," says a wistful engineer. "After all, I've got my Master's and now I have five years of engineering experience to boot."

Wistful could be talking about running down to the grocery store to pick up a carton of eggs. What Wistful doesn't know is that only approximately 20% of those with Master's degrees have the intellect and the motivation (or as one professor put it "the wheels and the drive") to complete a Ph.D. in aerospace engineering.

The first major step of becoming a Ph.D. candidate requires that you pass a set of grueling qualifying examinations. These involve up to six hours of written exams in advanced courses in astrodynamics and spacecraft control, hypersonic aerodynamics and entry theory, rocket propulsion and thermodynamics, and space structures. You may also be subjected to three hours of oral examinations by members of the faculty, who will ask you pointed questions about their areas of specialization. Some students faint dead away when confronted by the author of the famous textbook on *Principles of Dynamics*.

The process is mentally, physically, and emotionally taxing. No one who has gone through the qualifying exam will ever forget the experience or the mistakes they made.

One student had determined that the astrodynamics professor, Professor B, never asked anything about orbit decay. During the past five years, the professor had only asked questions about orbit transfer maneuvers high above the atmosphere. The student was exhausted from his preparations for the qualifying exam, so he elected to skip studying the theory of orbit decay due to atmospheric drag.

On the day of the qualifying exam the student turned to the astrodynamics question and guess what? It was a question about orbit decay!

With an impending sense of doom the student racked his brain trying to recall how to solve the problem. But he could not remember how to do it.

Getting his Ph.D. was the most important thing in life for this student, and now he was about to flunk the qualifying exam. Suddenly, the student had an inspiration. He decided to solve the problem as an orbit transfer problem. He could approximate the effect of the atmospheric flythrough as if the spacecraft had performed a retrobraking maneuver. It was a desperate act and he made some errors along the way.

Later the student was told that he had, in fact, passed the exam. However, Professor B said, "I want to see you in my office tomorrow morning."

That night the student looked up the problem he had not studied. He asked his friend, a brilliant student from Taiwan who had passed with flying colors the previous year, if he thought his new approach had any merit. The friend thought for a moment and said, "Yes, I think your idea will work—you should get points for it."

Encouraged, the student reexamined his new method and derived the equations carefully. The next morning he knocked at Professor B's door.

Professor B said, "You took a rather unorthodox approach in your exam—perhaps you'd like to explain yourself?" The student went to the professor's blackboard and picked up a piece of chalk. "This is what I should have done to solve your orbit decay problem." He wrote down the equations from memory. "It is based on the work-energy principle. And here is the formula for finding how the orbit decays with each atmospheric pass."

Professor B nodded his approval.

The student wrote some more equations. "This is how I approached the problem. It looks like an orbit transfer problem. My method is based on the principle of linear impulse and linear momentum. As you can see, Professor B, my derivation is shorter than yours and it gives the same answer!"

Professor B leaned forward and challenged: "But it can't have the correct sign!"

The student replied, "No, you see here—it is even the correct sign. The results are identical."

"Oh yes, I see that now," said the professor as he leaned back in his chair.

Suddenly the professor leaned forward again and said sternly "But you wouldn't teach it that way!"

To this the student acceded. His heart was pounding hard by now, but he felt vindicated, proud, and relieved all at the same time. He had survived the exam.

Years later the student, a doctor of aerospace engineering, inquired about Professor B and told the story of that "nearly fateful day when I took the qualifying exam" to a recent graduate of the school. The new graduate exclaimed, "But that's how he teaches it now!"

So the moral of the story is that the qualifying exam is a memorable experience—to say the least—and may even be the mother of invention. It is the most difficult exam you will ever take.

It is used by the faculty to weed out students who don't have the "Right Stuff" to become rocket scientists. After you pass the qualifying exam, however, getting your Ph.D. is a breeze. Although the flunk-out rate for the exam is 50 to 80%, the success rate in completing the Ph.D. after passing the exam is approximately 80%.

Chapter 29:
Why Working
on Your Ph.D. is Fun

There is a big difference that most students are unaware of between undergraduate school and graduate school. Undergraduate school can be an arduous experience. Most students take five or six courses per semester for four years. Some of the courses are dull, boring, or mind-numbing, and few have anything to do with rocket science. It is only in your junior and senior years that you get interesting courses in your chosen field. (Note: While these perceptions are typical of the engineering student, their reportage is not meant to indict general education, which is an integral part of the university experience.)

By the time a bright, motivated student finishes his B.S. in aerospace engineering, he is on the brink of burnout.

"I'm tired of school. I want to get into the real world and start making money. I don't think I could tolerate graduate school next semester."

Unfortunately, these are incorrect conclusions.

Graduate school is far more interesting than undergraduate school and involves fewer courses. You will take an average of only three courses per semester, approximately half the undergraduate load. The majority of your courses will be in your specialty. There are very few required courses. You will have more time to concentrate on fewer homework assignments.

You won't have to worry about money because most graduate students are supported by a fellowship, research assistantship, or teaching assistantship. You won't need a part-time job to pay for school.

The further you go in graduate school, the more focused your work becomes. Finally you are working on only one problem—your doctoral thesis—and no courses.

Working on your Ph.D. can be the most rewarding period of your life. It is research heaven. You have no financial worries, no family responsibilities, no mortgage—just the thesis.

That's why working on your Ph.D. is fun!

Even professors don't have it so good.

Chapter 30:
Plan Your Academic
Career Early

If you want to become a professor, you must plan your academic career from the day you start graduate school. Choose your advisor wisely and let him know immediately that you are planning to pursue a faculty position.

A Ph.D. is not enough! (Read *A Ph.D. Is Not Enough!* by Peter Feibelman; it's excellent!)

Choose an advisor who has produced professors among his students. Choose an advisor who is publishing strong research papers. Avoid large research groups. A professor who is directing a dozen doctoral students is probably not doing deep work. If your name is the thirteenth on a journal article, you will not have contributed much to research and the paper will not contribute much to your academic career.

You will need to publish several papers before you graduate with your Ph.D. In the old days, Ph.D. students never published journal articles. They published their theses first and then they published one or two journal papers after graduation. Getting a faculty position was easier then. If you had a Ph.D. from a good university and your advisor wrote a nice letter of recommendation, you could get the position.

Today you need three or four publications plus several manuscripts in preparation. You should be one of not more than four authors, and you should be the first or second author on half of the papers. Your publication record gives you academic visibility. Without this publication record, you don't have a chance of getting an interview at even a second-tier university.

Your Ph.D. from an outstanding university and your publications get you the interview. However, after you arrive for your interview you must also be able to answer that most frequently asked and often dreaded question: "How will you fund your research?"

Chapter 31:
How Will You
Fund Your Research?

During your interview for a faculty position you will be asked this question by the chairman of the department, the dean of engineering, and (third time's a charm) the provost of the university.

Suddenly, everything is about money. You may give wonderful answers about your technical research and the classes you would enjoy teaching, but if you cannot answer the money question you flunk the interview.

You must plan for this question far ahead of time. Just as you planned to write journal articles from the day you began graduate school, you must develop a plan to fund your research.

Take an interest in where your advisor is getting his funding. If you are being paid as a research assistant, where is the money coming from? Where will the money come from next year? Your advisor will be busy making funding contacts and writing proposals for funded research. Ask him about it and offer to write part of the proposal.

Your advisor will fall out of his chair from shock. When he recovers, he will have an expression of joy and wonder on his face. No graduate student has ever volunteered to help him write proposals before! He will gladly accept your help. Do an excellent job and he will have new respect and admiration for you.

This will not be lost when he is asked for a recommendation for your academic application. He will be able to honestly say that you were instrumental in obtaining funding while you were a graduate student and that he has no doubt that you will be able to fund your own research work.

The usual funding sources are NSF, AFOSR, ONR, ARO, and NASA. You can check their websites to find their specific research interests and who is in charge of those interests.

By the time you prepare for an interview, you should have contacted those individuals personally. A telephone call is required. A visit to the directorate is even better. Let your contacts know that you will be starting your academic career next year and want to get a head start writing proposals. They will be impressed and happy to talk to you.

It is often difficult for rocket scientists to establish funding contacts because of their tendency to focus on analysis rather than on people skills. Rocket scientists may be brainy, but they're also geeky. They prefer the pursuit of scholarship over dollarship. They resist trying to sell themselves and their research.

You may consider hitting the pavement in search of funding to be beneath your dignity. You may be angry at the system. You may be afraid of rejection.

You are not alone. But if you want to be a success as a professor of rocket science, you must overcome these obstacles.

Think of research funding as another type of scholarship. The directorates at NSF, NASA, and the military want to pay for research that will benefit their missions. They want the best researchers they can get. Your funding award indicates that the funding agency thinks you and your work are tops—you've won the competition.

If you want to win, you must break through your resistance, set aside your anger, overcome your fears. You can achieve this goal. You can create a funded research program. You have what it takes to plan the research and perform the work.

You may need to develop some new skills to sell yourself.

Ask your advisor for his help. He is an expert at raising funding or he wouldn't be your advisor. Keep in mind that the funding agencies are set up for the express purpose of doling out money to researchers like you. There are people waiting for their phones to ring, hoping to meet you—looking for an excuse to have lunch.

Arrange a business lunch and go. But don't call your contact between 11 a.m. and 2 p.m.—he won't be there—he's having lunch with your competitor!

You must establish personal contact. After all, if you were the funding agent and had to chose between two outstanding candidates from prestigious universities with visionary proposals, who would you trust with a $200,000 grant? A person you've never met or a pleasant, well-dressed, punctual person that you've met for lunch and with whom you've had many inspiring telephone conversations? Obviously you would choose the person you know over someone you've never met.

If you have taken all these considerations into account and prepared yourself for an academic career, you will be granted an interview for a faculty position.

Getting the invitation is a lot harder than you might think. For every faculty position advertised, there are usually more than one hundred, and often several hundred, applicants. To be granted an interview means you are among three to five candidates remaining out of hundreds who have applied. You get this far on the strength of your application and on the recommendations of your references.

You should have three documents in your application (besides the cover letter). They are: your resume, a list of courses you could teach, and your research plan.

Your research plan is of critical importance. (The list of courses and academic resume will be discussed later.) In it you should specify two to four research projects you will undertake as a faculty member. One of the projects can be an extension of your doctoral work. The other projects should contain new ideas. Write a single-page description of each project. Briefly explain the origin of the problem, the significance of the work, and what you expect the results will be.

Most importantly, give the names of the funding agencies that would be willing to support your research. Your research plan will have a great deal more credibility if you can name names of the funding agents themselves.

"I discussed this project with Dr. Zachary Smith, director of space propulsion in AFSOR during my visit in January of this year. Dr. Smith strongly encouraged that I submit a full proposal to him."

Your research plan, written down on paper and orally expressed at your interview, gives a resounding answer to the question, "How will you fund your research?"

Chapter 32:
What Should be on Your Academic Resume

The academic resume goes far beyond the resume for a general industry position. Your academic resume should include all publications, teaching experience, and work on funded projects.

You should list your journal publications (both in print and accepted for publication), your submitted journal articles, and your conference papers. You may wish to include a list of manuscripts in preparation, particularly if the publication list is short. But only do this if, in fact, you are actually working on them. Don't bluff—you may be asked to supply copies of these documents.

In your section on teaching, list and describe courses you taught or assisted teaching as a graduate student. Include student evaluations of your teaching, both numerical and written (such as excerpts of student comments). If you are a graduate teaching assistant with hopes of an academic career, be sure to get feedback on your teaching.

Describe funded research projects you worked on and your role in obtaining funding. If you helped your advisor write a funded proposal, you can list yourself as a coauthor (with your advisor's permission, of course). Give the name of the funding agency, the dollar amount of the grant, and the dates the project began and ended. Describe the significance and major results of your research.

Finally, provide a list of references. Do not be coy with "references provided upon request." The search committee doesn't have time for games—your resume will be tossed in the circular file, and you will receive your "ding" letter in the mail. Give a list of four outstanding references. At least three should be professors; the fourth could be a person in industry. Give full names, full titles, institution, full address, phones numbers (voice, secretary, and fax), and e-mail address. You want it to be easy for the search committee to make contact with your references.

Be very cautious in choosing your references. You will, of course, be expected to provide your advisor's name at the top of your list. If he does not appear, the search committee will assume you've had a falling out and may contact him anyway. If you were careful when you chose your advisor in the first place, you will have no trouble proudly listing him.

All your references should be people who will give bubbly, enthusiastic endorsements in support of your candidacy as a faculty member. Do not choose a "sad sack," a "wall flower," or a "critic" as your reference, regardless of how famous he is. Name recognition is important among your references (that's why you went to the best graduate school), but a lukewarm reference is the kiss of death even if it is from (or perhaps especially if it is from) a world-renowned researcher.

It is far better to have a less recognized, but extremely vocal champion of your cause. Pick only those references that you implicitly trust will give you the very highest recommendation. By the time you finish graduate school, you will know who they are.

If you have any doubt about your final selection, go to your potential reference and ask, "Professor Wizen, do you feel you know me well enough to give me a strong recommendation for a faculty position?"

Note that this question gives Professor Wizen an out, a way of saving his face and yours. Maybe he doesn't know you well enough. If he shows the slightest hesitation, drop him from your list.

If Professor Wizen asks, "Who are you?" then count your blessings that you discovered his amnesia regarding you and your work before you listed him as a reference. Having your reference tell the chairman of the search committee that they don't know you is extremely damaging to your chances of being hired.

Don't laugh—it has happened.

Be very cautious about who you entrust with your career.

3-4 one page research proposals . . .

Chapter 33:
List the Courses
You Could Teach

When you apply for a faculty position, you should submit three documents:
1) your academic resume,
2) your research plan, and
3) a partial list of courses you could teach.

In your list of courses you could teach, provide each course name and a description. Include separate sections on undergraduate and graduate courses. You can use your university's course catalog as a guide for the descriptions.

So what courses should you list? Anything you took as a student in your major area. Look at your transcripts and write down about a half dozen courses (or more) for each category—graduate and undergraduate. Only list the courses you would really want to teach.

You may also wish to develop new undergraduate- and graduate-level courses based on your research work or based on innovative teaching methods you'd like to try. Keep in mind that your teaching should reflect your research. Students will assume that what you teach is what you do. It's best for you to stay as close as possible to your research interests when you teach.

When applying for an academic position, look for weaknesses in the department that need to be strengthened. You must look carefully for clues indicating which needs must be fulfilled. Often, these needs are not clear in the advertisement for the position.

At some point you will talk to the chairman of the department. Listen attentively for any hints. If the chairman asks you if you'd be interested in teaching the Space Systems Design course, let him know in no uncertain terms, "I'd be delighted to teach such a course!"

Any answer less enthusiastic will lessen your chances of getting the position. You should be enthusiastic no matter what the course title is. Remember that course content is largely determined by the professor teaching the course. You can modify any course to fit your interests, so feel free to show your eagerness to teach whatever needs teaching.

If you aren't interested in teaching, then don't apply for an academic position. If all you really want to do is research, then go to a government lab or think tank. There are enough underachieving teachers in the university already.

When you send your teaching document in, it will contain an important course with exciting examples and innovative techniques. The course will be entitled "Space Systems Design." Just what the doctor ordered.

Anyone who ignores the casual suggestions of the chairman concerning the department's teaching needs will lose the position. You can verify a curriculum weakness by examining the department's catalog of courses or by visiting their website.

Point out the weakness in your cover letter and indicate how you would cure it. The faculty may not accept your proposed changes, but they will be very impressed with your keen interest in strengthening their curriculum.

If you do your homework (and follow the recommendations in this book), your three documents (resume, research plan, and list of courses) will put you head and shoulders above your competition. You will get an interview.

Whether you get the offer depends on how you prepare for and perform during the interview itself. At this point you look great on paper. But do you look great in person?

Chapter 34:
How Not to Give an
Academic Interview

The most important thing most faculty candidates fail to realize is that the department is not interested in your research.

So if you were expecting to profess long and hard about the great results you obtained in your thesis—and have done nothing else to prepare for your interview—then you are doomed to failure. The fact that you are getting an interview means that your resume fits the requirements of the department. You have a Ph.D., several published papers, and the right academic and research background.

You've stepped over hundreds of bodies to get this far. You are now one of three to five candidates. The interview will determine the winner, if indeed there is a winner. Often enough the department will reject all candidates they interview and start again next year.

The reason for rejection? Most candidates shoot themselves in the foot. The search committee is usually optimistic. They hope the candidate will do his job. They hope he will not shoot himself in the foot.

But four out of five candidates do it every time. The search committee shakes its collective head over the disaster. They don't want to extend the search for another year. If the last candidate could just work out—but there he goes—shooting himself in the foot. The committee is crestfallen when this happens.

Wasn't their ad clear enough? Didn't the candidate talk to his advisor? Didn't the chairman mention they needed someone to teach the design course? Didn't the candidate examine the department's website?

They shake their heads in wonder. Why aren't faculty candidates better prepared?

The answer to this seemingly perplexing question is simple: the candidates did not listen. They did not pay careful attention to the ad, the chairman, their advisor, or the website.

Why didn't they listen?

Because they are working on their Ph.D. thesis, the be-all and end-all to the theory of ballistic entry into planetary atmospheres. What could be more important than that? After all, this young person is brilliant. He's finishing his Ph.D. in record time. He's published in prestigious international journals. He's received the most prized scholarships.

Everyone—even he—knows he's a wonder.

And now Wonder Boy is going to give a no-holds-barred seminar on his Earth-shattering analytic theory that required one thousand pages of algebra to derive, but which he's cutting down to only 50 slides of equations, culminating in the grand equation of them all that has 150 terms and barely fits on three slides.

Wonder Boy is sure he will impress the search committee. He says to himself, "If their offer is sweet enough I might take it. I just hope the faculty is smart enough to appreciate my work. They better not expect me to teach much because I'll be busy extending my theory and if they think I should prostitute myself chasing money—well, I'll be out of there next year."

Wonder Boy proceeds to give a perfectly awful seminar.

Strike one.

During lunch with the faculty, the chairman asks Wonder Boy if he has any ideas about improving the curriculum. The ad specifically mentioned that the department wanted to strengthen its astronautics program. The chairman was hoping that Wonder Boy would say something about the Space Systems Design course.

Wonder Boy shoves another spoonful of chocolate mousse into his mouth. The faculty can barely hear his muffled reply.

"Oh—em—ya. I think your curr'clum's fine. Waterful. Wouldn't change a thing."

His words are garbled, but his message comes through loud and clear. He hadn't given a minute's thought about improving the curriculum.

Strike two.

Later, Wonder Boy is introduced to the dean of the engineering departments. The dean asks "How are you going to establish a funded research program?"

Wonder Boy thinks to himself, "Oh no—all these guys think about is money." He answers with a sneer, "Well, I think that scholarship is more important than dollarship, but I'm sure that NASA or somebody would be happy to handle my support."

Strike three.

Later, during small talk with the dean, the dean shows Wonder Boy a copy of his book, *The Aerodynamics of Insect Flight.* "Have you seen my book?" asks the dean.

"Never heard of it," replies Wonder Boy.

Nor does Wonder Boy show any interest in even looking at the book.

Strike four.

Advice to Rocket Scientists

Chapter 35:
How to Prepare for an
Academic Interview

So how do you prepare for an academic interview?

Do the opposite of everything that Wonder Boy did. Study the ad. Read between the lines. Call the department chair and ask for clarification. Listen intensely to the answer. Ask for a copy of the annual faculty report and the catalog of undergraduate and graduate courses. Check the website.

Study the curriculum. Look for weaknesses that you can fill. Make a specific plan for filling the weakness. This means writing an outline for one or two courses you could teach (for example, one graduate and one undergraduate course).

Study the faculty. Learn their names and their interests. Make special note of who won the best teacher award, who won the best paper award, who has an academic chair, and who brings in the most funding.

When you talk to these individuals, find a common ground of interest. Talk to their strengths. Avoid antagonizing the interviewer. If you feel that your publication record is really strong, then don't toot this horn when you're talking to the best teacher in the department who has a weak publication record. Ask instead, "What's your secret? How do you consistently win the best teacher award year after year?" The interviewer will be tickled to answer, and you'll learn important tips about teaching. You will make a friend.

When you talk to the most prolific researcher, let him know that you hope to follow in his footsteps. You have a good start in publications and want to continue. Ask, "Do you have any suggestions how I can maintain my momentum or even accelerate my production?" Again you've made another friend.

Or at least not an enemy.

Often a search committee will fall into two camps and fight over the last standing candidates. The division is usually over area of specialization: structures versus control, dynamics versus propulsion, aeronautics versus astronautics.

You have no control over this debate. You cannot change your stripes. If you are an astrodynamicist, everyone knows that you are not in space structures.

During your investigation you may discover what the two camps are. Or you may not. During your interview, you must demonstrate your strengths to those who would support you. Give them all the ammunition they need to argue your case.

Also, you must not antagonize the other side. During your interview, you must show everyone that you are a competent, decent, open-minded person with real drive and interest in contributing to the department. If the worst the opposing camp can say is, "He's a nice guy but he's an astrodynamicist!" then you've done all you can to maximize your chances.

You may have some difficult interviews. One particular difficulty occurs when you meet an individual with whom you have no common interests. Being locked in his office for half an hour seems like an eternity. Look around. Maybe you'll discover that he's purchased that new computer you're thinking of getting, or he's interested in your favorite sport, fly-fishing or whatever, or he reads *Sky and Telescope*.

If you find a topic, grab onto it and let him know you are interested. Even if you talk about the weather, it's better than staring at each other. Be genuinely friendly. Keep in mind that if you get the offer these will be your colleagues for years to come. Perhaps thirty years. Joining a faculty is more like becoming part of a family than getting a job. It's important that the faculty members and you feel that you can get along over the long haul.

As you meet each member of the faculty, think: "This may be my colleague. I may know this person for decades."

Give a firm handshake, look into his eyes—smile, and call him by his first name. If you are a fresh Ph.D. graduate you may find it difficult to call professors by their first names, but it is correct and necessary that you do so. You must break out of the shell of being a Ph.D. student. It would not be appropriate to address your new colleagues as "professor."

Pay careful attention to what your new colleagues say. Learn everyone's name and everyone's strength. Bring plenty of extra copies of your three documents (resume, course list, and research plan).

Find out what the department needs. (You should have a pretty good idea from your preinterview investigation.) Be clear to the faculty on how you would fulfill those needs.

If you find out that there is an introductory sophomore-level statics and dynamics class that has 100 students in it every semester and that the only professor who teaches it really needs some help with it, say, " I would be delighted to teach statics and dynamics." If you learn that the department really needs someone to teach space systems design because the professor who teaches it will retire next year, say "I would be delighted to teach space systems design."

If you are not delighted to teach virtually any course in your area at any level (especially the lower levels), then don't even apply for a faculty position. If you don't want to teach, then you don't want to be a professor. However, keep in mind that when you say, "I'd be delighted" you mean, "I'd be delighted to teach it my way." There is a lot of latitude in how you teach. You will be able to make the course your own. This is your advantage and your prerogative as a professor.

Remember that you are on a campaign. You want to prove to the faculty that you are the best person for their position. You must convince them one person at a time. When you meet each faculty member, say, "I'm very glad to meet you!" And mean it.

By the time you meet the dean, you have learned much about the aerospace department. You will be able to answer all of his questions.

"How will you obtain funding to carry on your research work?" Your answer: "I have prepared a research plan for you." Pull out your copy from your well-organized briefcase, hand it to the dean, and touch on the high points.

"What courses would you like to teach?" Your answer: "I have prepared a partial list of courses I could teach—here it is." Hand your list to the dean and add: "But I've learned that the department has strong needs in two courses: the sophomore statics and dynamics class and the space systems design class. I'd be delighted to help out with those courses as well."

"What do you think of your colleagues?" Be ready to rattle off the names of each faculty member in your area, a brief synopsis of what he or she does, and how you would collaborate with them in research or in teaching.

If you answer these questions well, you will practically have a lock on the offer.

There are a few pitfalls remaining. The greatest land mine of them all is the seminar. (More on this later.)

One more thing: if the dean shows you a copy of his recently published textbook on *The Aerodynamics of Insect Flight*, take it carefully into your hands (as if it were a holy relic). Page through it reverently. Show great interest and enthusiasm for it. Holding it in both hands, reluctantly give it back to the dean and say, "It's a beautiful book. I hope someday to accomplish such an undertaking. Congratulations!"

Do you really have to be told not to insult the dean? He is, after all, the one who gives final approval on all faculty hires.

Chapter 36:
The Academic
Seminar for Hire

Y ou've given great interviews. You've made friends. You impressed the chairman of the department and the dean of the engineering colleges.

But you can still screw up, because you have yet to give your seminar. The Seminar.

The faculty is waiting expectantly. They hope to hear a great seminar. If you haven't prepared the best seminar you've ever given in your life, then it will be easy to disappoint them.

There are many pitfalls. One of the biggest mistakes is to assume that the academic seminar for hire is like any other seminar or conference presentation.

It is not.

Unlike any other lecture you will give, this one requires that you actually lose your audience at some point. You must go over their heads.

You must demonstrate that you know far more about your subject than they can comprehend or ever hope to comprehend. You must break your audience like a bucking bronco tamed to ride. You must get them to submit that you are the undisputed Master of This Subject, that your work is World Class, that you are The Rising Star in This Field.

One of God's gifts to the intellectual world. A prima donna. After all, you are joining a club of prima donnas—that's what faculty members are.

If you aren't the Top Gunslinger in your neck of the woods, you will not be invited to join the club.

You are Not Deep.

No, if this faculty can follow everything you say in your area of specialization then you are shallow.

You will not impress any faculty if they understand your seminar completely. This is the paradox of the academic seminar for hire. You are not trying to clearly explain everything you've done.

But you must be careful, very careful about when you lose them. In this case, timing is everything.

You start out at the ground floor with great simplicity and clarity, but you gradually rise to higher and higher altitudes of abstraction and complexity that will have them gasping for air until they are ready to give up the ghost. Then, near the end of your seminar you plummet from the rarified heights down to breathable air. You resuscitate your audience with some numerical results that they can understand.

They are happy to be breathing again!

Then you give your list of conclusions. They are quietly stated but profound. Your audience is grateful. They've been taken on a ride to dizzying heights and have been returned safely to the Earth. They respect you.

A word of caution: remember that the timing of when you leave your audience behind and rocket skyward is crucial. It should not be too soon. You start your seminar like any other lecture (see Chapter 22 on how to give a presentation to rocket scientists and Chapter 23 on how to keep your presentation short and snappy). Begin with a highly simplified picture or cartoon. Set the stage. Discuss the history of the problem and its significance.

Make sure that everyone in the room understands what the problem is and why it is important.

The first third of your seminar should be crystal clear to the entire audience. The crucial rise into the stratosphere should occur in the second third of your lecture. No one should be left breathing when you complete the second third. In the last third of your lecture you bring them back alive.

Finish your one-hour lecture in 45 minutes. Do not go over or under this time limit. If you go over, a stampede of students will burst into the classroom—destroying your final effect. If the seminar is too short, the faculty will wonder if you ran out of material. Forty-five minutes is just right.

If you do everything right, the faculty will breathe a sigh of relief. They will leap to their feet and congratulate you on a job well done. You will know from their reactions that you've won them over.

You will get the offer.

Chapter 37:
Expect a Long Wait

After the interview you will have a long wait, an almost interminable one. It is best not to be in a hurry when dealing with any academic matters.

In the movie *Educating Rita*, Rita is worried that her professor (played brilliantly by Michael Caine) will be fired after delivering his poetry and literature lecture in an obviously inebriated condition. Caine looks up from his desk at Rita and says rather nonchalantly, "That would involve making a decision. They haven't done that for years."

So don't expect a phone call the week you get back from the interview. The professors have to meet. They have to discuss the merits of each candidate. They have to vote. What could be more complicated? Or time consuming?

If you are lucky, you will be contacted in a month. It could easily take three months, however, and if the search committee deadlocks you may never hear from them.

In the latter case, the chairman may finally call you and let you know that no decision has been made, which is in fact a kind of decision—the decision not to hire anyone that year.

The deadlock or decision not to hire happens fairly often, approximately 25% of the time. Do not be discouraged. Do not be in a hurry. Take a postdoctorate or an industry position and apply for positions at other universities next year. Keep building your resume. If you were interviewed, it means your resume is in good shape.

Carefully review what went on during the interviews. You will probably be painfully aware of the embarrassing moment that lost the position for you. If that's the case, you know what to work on for your next interview.

Sometimes a rejection is simply a matter of not fitting in because of professional or personal reasons. If you didn't feel comfortable with the faculty, then they didn't feel comfortable with you. (You will get this feeling very early in the interview process, if at all.)

Sometimes a rejection is because of a hidden department agenda. You may never know. If you are rejected, but have made a friend in the department, you may wish to call him to find out what went wrong. The answer may or may not be of benefit to you.

If you come up with a complete blank as to why you were rejected, ask a straight-shooting close friend what he thinks. Perhaps you have halitosis or bad table manners. It is painful to be told this, but personal hygiene and manners matter. You can fix the problem and go on.

You may have to go through several interviews, perhaps half a dozen or more before you get an offer. Consider each interview an "at bat." The batter cannot expect to get a hit every time he steps up to the plate, but for each trip he's ready. He takes his best stance and swings hard at his pitch. If he gets a hit for every third at bat, he's a star player.

So, take your at bats. If you keep applying the principles in this book you will finally get an offer. The phone will ring with good news.

Chapter 38:
How to Negotiate
an Academic Offer

When you get the call from the dean or chairman offering you the position, you must be ready to start negotiations for the specific conditions of your accepting the offer.

You must prepare your demands in advance.

A typical offer will include the following:

1) The going rate for an academic salary for a person of your experience. This salary will be for nine months per year so it will be less than yearly salaries offered in industry, but it should be proportional (approximately three-fourths of the industry rate). Do not accept a lower salary. Contrary to popular belief, professors do not make less than their colleagues in industry. In fact they can earn more, often substantially more, in their academic positions. Their salaries are largely dependent on the prestige and funding they bring to the university.

2) Summer salary for the first (and sometimes second) year. You should expect to receive at least two months' summer salary for your first year. After that you are expected to bring in your own funding to cover your summer pay. The idea of the summer salary offer is to help you through the transition period when you haven't yet brought in any grants.

3) Start-up funds. You can expect some funds from the department to help you get started. These funds help pay for your first graduate student, usually for one to two academic years. Without a research assistantship you will have difficulty getting graduate students to work on your projects. Your start-up funds also include money for a computer and peripherals. Be ready to specify the model numbers for the computer, printer, and other equipment, and know how much everything costs. If you are an experimentalist, you will need a detailed list of all equipment you require. During your interview, you should take careful notes of what equipment is already available and what new equipment you will need to purchase to get your experiments started.

When you receive the offer, you must be ready with your list of requirements. The dean and chairman will be in a hurry to seal the offer. They don't want to lose you because of delays, nor do they want to lose their second candidate if you drop out.

Know what you need and what it costs. Do not accept an offer that does not permit you to start your research program immediately.

You will never be in a better position to negotiate. The letter from the dean will spell out in writing what you are getting in the offer. Anything not in this letter is not guaranteed. Do not accept anything less than a written offer with all of your requirements specified.

If you are told, "Maybe next year we can get you that expensive computer you asked for," you can be sure you won't be getting it.

Even though it is very important to be clear about your requirements, you must exercise some flexibility in making your demands. Listen to the dean and the chairman. There may be limits on what they can offer in the way of salary. For example, some state universities have rigidly set salary levels for different levels of experience. However, the dean may be able to offer you more funding for graduate students or for equipment to compensate for the lower salary.

In other cases, you may be able to get a better salary offer. This is especially true if you have more than one academic position offered to you.

Keep in mind that the dean and chairman are highly motivated to hire you. They will do everything in their power to meet your demands. But when they mention limits, you should listen carefully. You won't be the only one disappointed if the deal falls through—so will the dean, the chairman, and the faculty.

The most important thing in negotiations is to keep the lines of communication open. Keep a positive attitude and a sense of humor. Know what you need and what you can accept.

If you follow these guidelines you will come to a mutually beneficial agreement. You will receive an offer that assures you the best chances of success in your academic career.

$$\vec{F} = m^{e}\ddot{\vec{r}}^{oc}$$

Part VI

How to Get Promoted in Academia

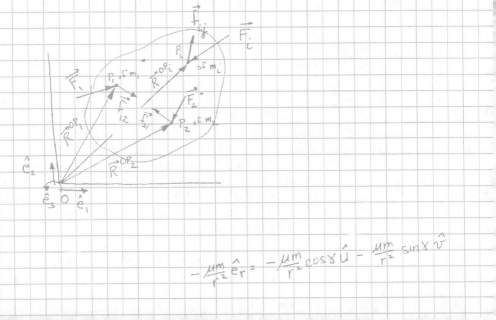

$$-\frac{\mu m}{r^2}\hat{e}_r = -\frac{\mu m}{r^2}\cos\gamma\,\hat{u} - \frac{\mu m}{r^2}\sin\gamma\,\hat{v}$$

Chapter 39:
What it Takes to Get Tenure

The academic ladder has three rungs: assistant professor, associate professor, and (full) professor. But rank is nothing if you don't achieve tenure within a prescribed time limit.

In the typical case, a fresh Ph.D. joins the faculty as an assistant professor and has six years to make tenure. If in that time he does not attain tenure, he will be given one more year of employment; that is, he will be allowed to finish his seventh year, at which point he must leave the university.

"Didn't get tenure" is a stigma that will follow you for the rest of your life. It's like the mark that God put on Cain.

If you don't achieve tenure, no other top tier university will touch you. You may not even be welcome in a second-tier school. (After all, they too have "standards.") You'll be forced to go to industry for a job. Everyone in the country will know what happened to you. The aerospace community is a small world.

And bad news travels fast. Not getting tenure is the death of an academic career.

How can you avoid such a disaster?

The official answer is that you must make significant contributions in three areas:

1) research,

2) teaching, and

3) service.

The unofficial answer is that you must make significant contributions to research, research, and research. The unofficial answer used to be right. Teaching didn't matter at all and service on committees mattered very little if the candidate published dozens of papers, graduated several Ph.D. students, and brought in tons of funding.

Lately, however, promotion committees have required that teaching must be more than lip service. You can no longer do a bad job of teaching and expect to get tenure.

Unfortunately, many young assistant professors mistakenly become great teachers and expect that this will gain them tenure. Six years later, after winning a string of best teacher awards, the young assistant professor is denied tenure.

A storm of protest is heard across the campus. "I don't understand what is wrong with this university," a student will say. "This is the best teacher I've ever known—and he's being fired."

The next year, Best Teacher packs his bags and leaves for a job in industry. What the students and public in general don't know is that Best Teacher never published a single paper, never obtained a research grant, never produced a Ph.D. student, and never served on a committee. He did only one third of the job and he did it very well!

Anyone can take one third of a job and do it well. What your must do is all three tasks—research, teaching, and service—with emphasis on research.

So what is service? Service means serving on committees. It is doing your part in administering the department's tasks. You have many committees to contribute to, including the curriculum committee, the computer committee, the graduate admissions committee, the design committee, the colloquium committee, the grades arbitration committee, the teaching awards committee, the library committee, the picnic committee, etc.

All the activities of the department are done by the professors with the support of a small staff. You must do your part. What you don't do will be noticed by the full professors on the promotion committee. What you don't do must be done by someone else.

A quick way to make enemies among the promotion committee is to shirk your service responsibilities. This infraction will not be forgotten. It will be held against you and your tenure may be delayed.

It is often said that tenure can be based on any one of the three tasks of research, teaching, and service. But in fact it is never granted for anything but exemplary research, adequate teaching, and adequate service.

So what must you do in research to achieve tenure? Three things: publish, obtain funding, and turn out Ph.D. students.

"How many papers must I publish to get tenure?" You will ask this question—everyone does—and you will get no answer. The promotion committee seems to be keeping it a secret. It seems they're afraid to give you a specific number—perhaps you'll sue them if you deliver and they don't promote you. Papers count and they count papers (they do little else—they cannot evaluate your work) but they won't tell you the magic number.

"There isn't a magic number," they say, "but we know a winner when we see one."

But rocket scientists are numbers people. You need a target. Here's one: fifteen. Publish fifteen full-length journal articles (not conference papers—they don't count) in several different, prestigious national and international journals. Twenty years ago the target was eight, but now (in 2002) it has risen to approximately fifteen.

Talk to your colleagues privately. Ask them how many papers they had at the time of their promotions. Take into account how recent their promotions were. Most of your colleagues will tell you.

You can learn a lot just by asking.

"How much funding is enough?"

You should bring in one or two mid-sized grants. You should have enough money to pay yourself for at least two or three summers and to support a graduate student for three years. If you are an experimentalist, you also need grants to purchase crucial equipment.

So you will need $100,000 to $250,000 (in 2002 dollars) in research grants. Essentially, you should establish a research program that is well funded and likely to continue to be well funded for the foreseeable future. If you have enough funds to always pay for your summer salary and the graduate students and equipment you need, then you have adequate funding.

If you're feeling financially pinched because you are not drawing funds to cover your own summer salary, or if you find yourself writing all your papers because you cannot afford a graduate student to help you, then you do not have sufficient funding to be granted tenure. Simple as that.

Of course, if you do not hire graduate students, you will not graduate a Ph.D. student in time for your tenure. It takes three to five years for a graduate student to finish a doctoral degree and he will need support throughout that time. If you don't have the money, you won't get the students, and you won't have what it takes to get tenure.

Chapter 40:
Train Your Graduate
Students to do Research

Some assistant professors make the mistake of doing all the research themselves. They are often quiet, brilliant people who do their best work alone. If you follow this approach, you may have twenty full-length journal articles published, but no Ph.D. students will have graduated when your tenure decision comes up.

You must remember that professors teach. They teach at every level from freshman to postdoctorate. Teaching Ph.D. students to do research is a crucial part of the program. At first you may find explaining your work to a new Ph.D. student slows you down. Of course it does: it is a time investment. But it will pay large dividends in the future.

The department will give you one to two years of funding to support a graduate student. It is important that you use this funding wisely. Make sure you get something in return from your student.

Have your student give weekly stand-up presentations of his work. He should have five to fifteen slides and should make a hard copy for you. Do not let him report to you sitting down. Make him use the overhead projector.

To start, you can assign your student a paper to read—perhaps one of your own journal articles. Have him present his analysis of your paper over the next two or three weeks. It takes at least that long to duplicate the work. The student will then have a basis for extending the research.

While he gives his presentation, ask questions and give advice. Make him write down all your suggestions. Your student may have a 20-minute talk prepared for you, but your meeting with him will take an hour. At the end of the meeting, your student will have plenty to do for the next week.

This is a powerful technique for getting results out of your graduate students.

For some reason, the stand-up presentation with overhead slides brings out the best in student performance. It is much more difficult for the student to bluff about the work he did. When projected onto the overhead screen in large letters, the weaknesses and holes in his analysis are highly magnified.

Under no circumstances should you let your students give sit-down reports to you or meet with you on an irregular schedule. The weekly stand-up presentation is the most efficient way to get productive research from your students.

Of course, you will find out very quickly if your student is lazy or incompetent. He will not be able to make progress. His weak presentations will demonstrate that he's missing the wheels or the drive or both.

On the other hand, if your student is good he will quickly progress. You will have enough research material from him after one year to present a conference paper.

Take your student to the conference and have him make the presentation. This is good experience for him. The practice he had presenting to you will be evident in his confidence and clear thinking during his conference presentation.

When you return from your conference trip, have your student rewrite the conference paper for journal submission and send it out. If your student has contributed more than half of the good ideas in the paper and more than half of the writing, then put his name as the first author. Thereafter you can alternate first authorship with your student.

Keep in mind that it doesn't matter to the promotion committee if your name is first or last. They know that it was your funding and your ideas that brought the paper into existence. Regardless of where your name appears, it is assumed that you are the head honcho behind the research.

Your student will be galvanized into doing his best work when he sees that you always give him full credit for his effort. Placing his name first on all of the conference papers and many of the journal articles is an excellent way to reward him.

You can use this technique with several Ph.D. students simultaneously. Depending on your funding and ingenuity, you can advise one to six doctoral students who will produce one to six papers per year for you.

If you can support an average of three graduate students per year over your six-year tenure track, you will easily have everything you need (papers, funding, and Ph.D. graduates) to qualify for tenure.

Chapter 41:
How to Get Promoted and Tenured at a Higher Rank

If you start as an assistant professor, you will get promoted to associate professor at the same time you receive tenure. If you had experience before accepting an academic position you may be offered the rank of associate or full professor, usually without tenure. In these cases you will be given three years to achieve tenure if you start as associate and only one year to achieve tenure for full professor!

Your salary offer will be tied to your experience and not to your rank. If you are leaving industry to begin an academic career, you may wish to take the rank of assistant professor to have a comfortable six-year fuse before the tenure decision. If you achieve tenure early, then you have lost nothing. Your starting salary will be the same.

The standards for promotion from associate professor to full professor are similar: you do more of the same. More research, more teaching, and more service. You should attain national or international recognition as a leader in your specialty, and you should be considered a valued colleague in your department.

You want numbers? OK. You should have 35 full-length journal articles (not technical notes, not conference papers), five Ph.D. students graduated, and approximately $1 million dollars in funding (in 2002 dollars). In 1980, the number of papers was approximately 20, so these numbers continue to rise—it's a moving target. Strike as early as you can. After you get tenure and associate rank, don't take a break. Work hard and fast to get promoted to full professor.

Under rare circumstances, an individual from industry can enter the university as a full professor with tenure and at a very high salary. This person will have 50 journal articles or more, transferable funding, and a book or two under his belt. He is a world-renowned authority in his field.

But fifteen years of experience in industry without publications is not a ticket to academia. If you aren't publishing as least three journal articles per year, no university will have you. To get an academic appointment you must be publishing regularly.

The old saw—publish or perish—is alive and well.

Epilogue

I hope that some of the advice I give here has been of value to you. I am also very interested to learn about your experiences. Please write to me if you'd like to share your personal anecdotes and ideas. With your permission, I may include them in a future edition of *Advice to Rocket Scientists*.

About the Author

After receiving his Ph.D. in Aerospace Engineering from The University of Michigan in 1979, Jim Longuski (long-gŭs'-skē) worked at the Jet Propulsion Laboratory as a maneuver analyst and as a mission designer. In 1988 he joined the faculty of the School of Aeronautics & Astronautics at Purdue University in West Lafayette, Indiana, where he teaches courses in dynamics, aerospace optimization, and spacecraft design. He is coinventor of a "Method for Velocity Precision Pointing in Spin-Stabilized Spacecraft or Rockets" and is an Associate Fellow of AIAA. Professor Longuski has published over 100 conference and journal papers in the general area of astrodynamics including such topics as spacecraft dynamics and control, reentry theory, mission design, space trajectory optimization, and a new test of general relativity.

Recommended Reading

Arnot, R., *The Biology of Success,* Little, Brown and Company, Boston, 2000.

Augustine, N., *Augustine's Laws,* AIAA, Reston, VA, 1997.

Axelrod, A., *Patton on Leadership: Strategic Lessons for Corporate Warfare,* Prentice Hall Press, Paramus, NJ, 1999.

Berne, E., *Games People Play: The Psychology of Human Relationships,* Ballantine Books, New York, 1973.

Buckley, R., *Strictly Speaking: Reid Buckley's Handbook on Public Speaking,* McGraw Hill, New York, 1999.

Carnegie, D., *How to Win Friends and Influence People,* Pocket Books, New York, 1961.

Feibelman, P.J., *A Ph.D. Is Not Enough! A Guide to Survival in Science,* Addison-Wesley Publishing Company, Reading, MA, 1994.

Feynman, R.P., *"Surely You're Joking, Mr. Feynman!" Adventures of a Curious Character,* Bantam Books, New York, 1988.

Feynman, R.P., *"What Do You Care What Other People Think?" Further Adventures of a Curious Character,* W.W. Norton & Company, New York, 2001.

Flesch, R., *The Art of Clear Thinking,* Barnes & Noble Books, New York, 1973.

Fox, J.J., *How to Become CEO: The Rules for Rising to the Top of Any Organization,* Hyperion, New York, 1998.

Fox, J.J., *How to Become a Rainmaker: The People Who Get and Keep Customers,* Hyperion, New York, 2000.

Harris, T.A., *I'm OK–You're OK: A Practical Guide to Transactional Analysis,* Harper & Row Publishers, New York, 1969.

Harrison, A.F., and Bramson, R.M., *The Art of Thinking: Strategies for Asking Questions, Making Decisions, and Solving Problems,* The Berkley Publishing Group, New York, 1984.

Laird, D.A., and Laird, E.C., *Sizing Up People,* McGraw-Hill Book Company, New York, 1964.

Laird, D.A., and Laird, E.C., *The Technique of Getting Things Done: Rules for Directing Will Power from the Lives of the World's Leaders,* McGraw-Hill Book Company, New York, 1947.

LeBoeuf, M., *The Greatest Management Principle in the World,* Berkley Books, New York, 1989.

Morrell, M., and Capparell, S., *Shackleton's Way: Leadership Lessons from the Great Antarctic Explorer,* Viking Penguin, New York, 2001.

Peter, L.J., and Hull, R., *The Peter Principle: Why Things Always Go Wrong,* Bantam Books, New York, 1970.

Sheehy, G., *Passages: Predictable Crises of Adult Life,* Bantam Books, New York, 1977.

Stine, J.M., *Double Your Brain Power,* Barnes & Noble Books, New York, 2000.

St. James, E., *Simplify Your Life: 100 Ways to Slow Down and Enjoy the Things That Really Matter,* Hyperion, New York, 1994.

vos Savant, M., and Fleischer, L., *Brain Building,* Bantam Books, New York, 1991.

Zinsser, W., *On Writing Well,* Harper & Row Publishers, New York, 1988.

Zinsser, W., *Writing to Learn,* Harper & Row Publishers, New York, 1989.